Multiple Stressors: Risk-Based Framework and Experimental Design for Cause/Effect Relationships

by:

Stella M. Swanson, Ph.D.
Christian G. J. Fraikin, D.E.C.
David J. Van Den Assem, M.E.Des.
Sandra Vaz da Costa Vargas, Ph.D.
Golder Associates Ltd.

Joseph M. Culp, Ph.D.
Richard B. Lowell, Ph.D.
National Water Research Institute, Environment Canada

Susan Eipert, Ph.D.
Eipert Information Services

Gary A. Lamberti, Ph.D.
University of Notre Dame

Barry R. Taylor, Ph.D.
Taylor Mazier Associates

2004

WERF

The Water Environment Research Foundation, a not-for-profit organization, funds and manages water quality research for its subscribers through a diverse public-private partnership between municipal utilities, corporations, academia, industry, and the federal government. WERF subscribers include municipal and regional water and wastewater utilities, industrial corporations, environmental engineering firms, and others that share a commitment to cost-effective water quality solutions. WERF is dedicated to advancing science and technology addressing water quality issues as they impact water resources, the atmosphere, the lands, and quality of life.

For more information, contact:
Water Environment Research Foundation
635 Slaters Lane, Suite 300
Alexandria, VA 22314-1177
Tel: (703) 684-2470
Fax: (703) 299-0742
www.werf.org
werf@werf.org

This report was co-published by the following organization. For nonsubscriber sales information, contact:

IWA Publishing
Alliance House, 12 Caxton Street
London SW1H 0QS, United Kingdom
Tel: +44 (0) 20 7654 5500
Fax: +44 (0) 20 7654 5555
www.iwapublishing.com
publications@iwap.co.uk

Library of Congress Catalog Card Number: 2004103669

IWAP ISBN: 1-84339-707-2

Golder Associates Ltd., Taylor Mazier Associates, National Water Research Institute, Eipert Information Services, University of Notre Dame

The research on which this report is based was funded, in part, by the United States Environmental Protection Agency through Cooperative Agreement No. CR-827345-01-1 with the Water Environment Research Foundation (WERF). Unless an EPA logo appears on the cover, this report is a publication of WERF, not EPA. Funds awarded under the Cooperative Agreement cited above were not used for editorial services, reproduction, printing, or distribution

ACKNOWLEDGMENTS

Supplemental support from the U.S. Environmental Protection Agency is gratefully acknowledged. Members of the Project Subcommittee were involved in two planning sessions, providing lively discussion and valuable contributions to the direction and content of the report both in these sessions and throughout the duration of the project. Their contributions are greatly appreciated. The project team thanks Mr. David O'Sadnick and Ms. Shannon Poppe of Golder Associates in Denver for hosting the year-one planning session.

Report Preparation

Principal Investigator:
Stella M. Swanson, Ph.D.
Golder Associates Ltd.

Project Team:
Joseph M. Culp, Ph.D.
National Water Research Institute

Christian G.J. Fraikin, D.E.C.
Golder Associates Ltd.

Gary A. Lamberti, Ph.D.
University of Notre Dame

Richard B. Lowell, Ph.D.
National Water Research Institute

Geetha Ramesh, Ph.D.
Golder Associates Ltd.

Barry R. Taylor, Ph.D.
Taylor Mazier Associates

David J. Van Den Assem
Golder Associates Ltd.

Sandra Vaz da Costa Vargas, Ph.D.
Golder Associates Ltd.

Project Subcommittee

Scott Dyer, Ph.D., Chair
Procter & Gamble

Michael V. Bastian
CH2M HILL

Timothy J. Hall
National Council for Air and Stream Improvement, Inc.

Carolyn T. Hunsaker, D.Env.
U.S.D.A. Forest Service

Dixon H. Landers, Ph.D.
U.S. Environmental Protection Agency

Irwin Polls
Metropolitan Water Reclamation District of Greater Chicago

Kent W. Thornton, Ph.D.
FTN Associates

Water Environment Research Foundation Staff

Director of Research:	Daniel M. Woltering, Ph.D.
Project Managers:	Dean Carpenter, D.Env.
	Eugenia Bruno Hart

ABSTRACT AND BENEFITS

Abstract:

Confident prediction of multiple-stressor effects in aquatic ecosystems is central to confident decision making in water pollution control. This study team develops a flexible and scientifically defensible conceptual model of the environmental effects of multiple stressors on river and stream ecosystems. The model is then tested by a science-based study design.

The conceptual model includes two main components: 1) a diagram that portrays the main features of the study team's understanding of how multiple stressors would interact and the nature of the effects; and 2) three testable risk hypotheses. The model is consistent with the guidance provided by the U.S. EPA for ecological risk assessment (U.S. EPA, 1998).

The study design has an experimental emphasis because there is a lack of published experimental studies that test multiple-stressor hypotheses. The study design uses a weight of evidence assembled using the results of experimental mesocosm tests. The mesocosm tests are supported by experimental laboratory data and observational field data. Results of the experimental test will be used to refine the initial risk hypotheses and the conceptual model. The overall goal is to use repeated iterations of this test scheme to produce a robust and flexible conceptual model that can be used to predict ecosystem responses to multiple stressors in a wide variety of situations.

Benefits:

- Presents risk managers with a conceptual model and risk hypotheses that will improve the ability to confidently predict the effects of multiple stressors in lotic freshwater environments. This will allow for more cost-effective wastewater treatment by identifying the stressors that cause the greatest effects.
- Provides a study design for testing the risk hypotheses, including a General Test Scheme and criteria for selection of sites, stressors, receptors, and endpoints.
- Describes a weight-of-evidence approach for assembling the information and data evidence from laboratory, mesocosm, and field multiple-stressor studies.
- Proposes a new method focusing on an objective, concise, and transparent manner to assemble the weight of evidence.
- Includes a hypothetical step-by-step example that guides the reader on how this weight-of-evidence method can be used in practice.

Keywords: conceptual model, risk hypotheses, ecological risk assessment, mesocosm, weight of evidence, risk ranking, ecosystem response

TABLE OF CONTENTS

LIST OF TABLES

LIST OF FIGURES

LIST OF ACRONYMS

2,4-D	2,4-Dichlorophenoxyacetic Acid
ANOVA	Analysis of Variance
BOD	Biological Oxygen Demand
DO	Dissolved Oxygen
EPT	Ephemeroptera/Plecoptera/Trichoptera
LOEC	Lowest Observed Effect Concentration
NCASI	National Council for Air and Stream Improvement
NPS	Nonpoint Source
P/B Ratio	Production to Biomass Ratio
P/R Ratio	Production to Maintenance Ratio
PAH	Polycyclic Aromatic Hydrocarbons
PCB	Polychlorinated Biphenyls
PSC	Project Subcommittee
R/B Ratio	Maintenance to Biomass Ratio
SI	Stressor Identification
TMDL	Total Maximum Daily Loads
SETAC	Society of Environmental Toxicology and Chemistry
U.S. EPA	United States Environmental Protection Agency
WERF	Water Environment Research Foundation
WWTP	Wastewater Treatment Plant

EXECUTIVE SUMMARY

A general set of theories regarding the response of aquatic ecosystems to several sources of physical and chemical stress (multiple stressors) is required to provide a confident basis for prediction of multiple-stressor effects. Confident prediction is central to confident decision making in water pollution control.

The merits of developing a conceptual model for multiple stressors that can be experimentally tested and confirmed in the field were identified by the Water Environment Research Foundation (WERF) in 1999. WERF subsequently commissioned this project, "Multiple Stressors: Designing a Model for Risks and Relationships" (00-ECO-2), to provide a science-based study design that helps establish a general, conceptual approach to multiple stressors.

The objectives of this study are to:

- Review and critique the existing body of knowledge for multiple stressors
- Develop a searchable, annotated bibliography of multiple-stressor research (available upon request)
- Identify gaps in the body of knowledge
- Develop a flexible and scientifically defensible conceptual model of the environmental effects of multiple stressors on river and stream ecosystems
- Develop a study design to test the conceptual model of multiple-stressor effects
- Document the process used to arrive at the conceptual model

This report addresses the last three objectives. This report includes the conceptual model for multiple stressors, the study design to test the conceptual model, and a brief description of the process used to arrive at the conceptual model. The first three objectives are addressed in a companion report, "Multiple Stressors: Literature Review and Gap Analysis" (Swanson et al., 2004; WERF Report 00-ECO-2B).

ES.1 Developing the Conceptual Model

The multiple-stressor conceptual model includes two main components: 1) a diagram that portrays the main features of the study team's understanding of how multiple stressors would interact and the nature of the effects; and 2) a series of three testable risk hypotheses. The conceptual model provides a framework for understanding the interactions between stressors and their effects on habitat, structure, and function within an ecosystem, based on different baseline conditions. Baseline conditions refer to the existing ecological conditions in a particular habitat before the addition of any specific stressor(s).

The conceptual model addresses types of ecosystem response to stressors, relationships among stressors and responses, and the cumulative effects of stressors.

ES.2 Conceptual Model Diagram

The proposed conceptual model illustrated by the linkage diagram (Figure ES-1) encompasses the following key assumptions:

- Interactions between the stressor and the ecosystem depend on the existing baseline conditions.
- For each stressor/baseline combination there will theoretically be a functional and a structural response.
- Effects on habitat are caused by structural or functional responses to stressors.
- Effects on habitat can cause structural or functional responses.

Baseline conditions are important because ecosystem responses will be different depending upon the existing stressor concentrations. An undisturbed section of river with background concentrations of stressors will respond differently than a section of river that has received several point and nonpoint discharges upstream. The dotted lines in the linkage diagram illustrate the influence of baseline conditions on structural, functional, and habitat responses in the ecosystem.

Structural responses are responses in the make-up of the biological communities. For example, the species composition of the algae may change. Changes in communities can have a "ripple" effect through food chains and can cause changes in ecosystem function.

Functional responses are responses in the energy and nutrient cycling of ecosystems. For example, photosynthesis rates of algae may increase, leading to more rapid growth and reproduction and higher biomass. Functional responses can also have a ripple effect—causing functional changes up the food chain as well as shifts in community structure.

Habitat responses are fundamentally linked to ecosystem structure and function. The broad definition of habitat includes physical, chemical and biological characteristics required by organisms to successfully grow and reproduce. Direct physical alteration of habitat by human activities can cause structural and functional responses in ecosystems.

The solid lines in the linkage diagram illustrate the interrelated nature of structural, functional, and habitat responses to stressors.

The conceptual model diagram does not explain the expected sequence of responses, nor does it portray the expected interaction among multiple stressors. For example, would a particular combination of stressors produce a structural change first, and would the effect of the stressors be additive?

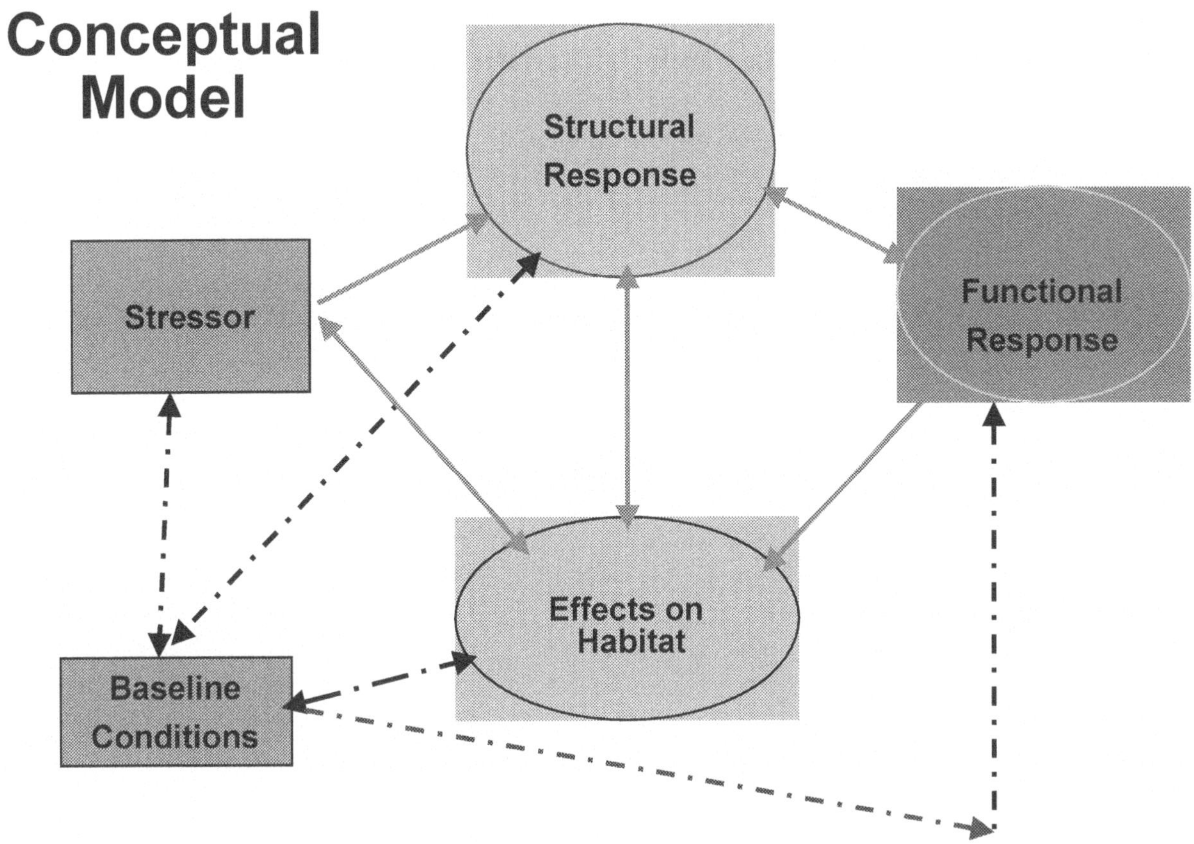

Figure ES-1. A linkage diagram of the conceptual model.

ES.3 The Risk Hypotheses

Risk hypotheses complete the conceptual understanding of multiple-stressor effects. The study team developed an overall theory of ecosystem response to multiple stressors based on the central notion that almost every stressor is potentially either a subsidy (producing stimulatory effects) or a toxicant (producing inhibitory effects) in an ecosystem, depending upon its concentration. The study team then proposed that all subsidy/toxicant response curves for different kinds of stressors had the same general shape; the difference lying where on the curve the new input places the ecosystem in question.

This is a new model for ecosystem response (although grounded in earlier theories).The model combines the subsidy/toxicant dichotomy into a single relationship, addresses a longstanding controversy about whether functional or structural changes are expected first for any given situation, and leads to interesting new hypotheses. Most important, this model is testable and can be used to predict ecosystem responses to different kinds of multiple stressors in different baseline conditions.

The above central notion led to the development of three risk hypotheses: 1) subsidy stressors first cause a functional response, 2) toxicant stressors first cause a structural response, and 3) multiple stressors will act in an additive manner.

The first hypothesis assumes an almost instantaneous functional response to subsidies. For example, heat produces a rapid increase in metabolism, or that the addition of a limiting nutrient produces a rapid response in primary production of algae.

The second hypothesis assumes that toxicants will change structure first because of the rapid reduction or elimination of sensitive species. This would be particularly true if acutely lethal thresholds were exceeded.

The third hypothesis assumes that when faced with two or more stressors, population declines will happen more swiftly, species replacements will be accelerated, and loss of species will begin at an earlier threshold of each stressor than for either stressor alone.

Synergistic or antagonistic responses are also possible but are likely to be less common than simple additive responses.

An additional important consideration for each of the above three hypotheses is that the responses to different stressors will invariably depend upon the existing baseline (or background level) conditions.

In addition to the linkage diagram and the risk hypotheses, the conceptual model includes the subsidy-toxicant curve (Figure ES-2), which shows that nearly all stressors can potentially be a subsidy or a toxicant to an ecosystem, depending on its dose or intensity.

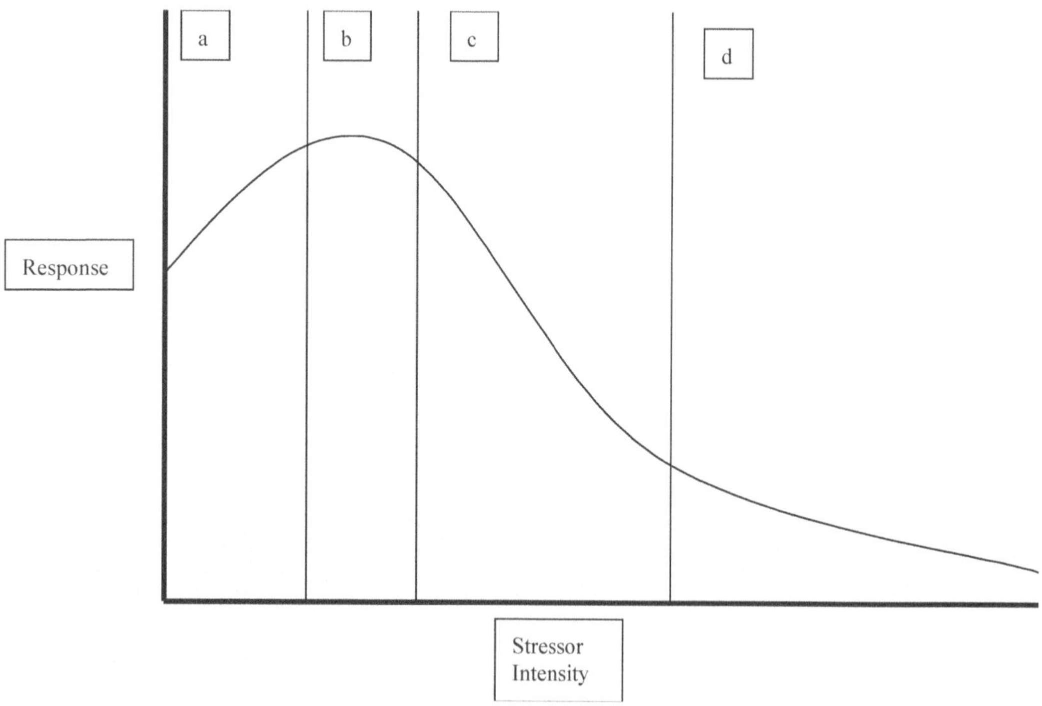

Figure ES-2. Subsidy-Toxicant Curve.

The different sections labeled [a], [b], [c], and [d] correspond to different concentration/response combinations, such that the positioning of the *y* axis would reflect whether a stressor's concentration/intensity is a subsidy or a toxicant to the ecosystem.

Quite different outcomes should be expected in situations where baseline conditions are already "disturbed," where toxicant concentrations are above baseline and/or the ecosystem is already nutrient rich. The *y* axis of the stressor-response curve would be shifted (as illustrated by the [b], [c], and [d] positions on the general stressor-response curve).

ES.4 The Study Design

The study design has an experimental emphasis because there is a significant lack of published experimental studies that test multiple-stressor hypotheses. The primary tool for testing the risk hypotheses is the outdoor mesocosm. Compared to single-species studies in the laboratory, community-level outdoor mesocosm experiments incorporate greater environmental realism than single-species laboratory experiments, are better able to identify components within an aquatic community most sensitive to a given stressor, and integrate a variety of ecological processes. Flowing-water mesocosm techniques can identify mechanisms of stressor effects, investigate cause/effect relationships, and eliminate confounding environmental factors that are common to field observational studies.

Under the integrated General Test Scheme (Figure ES-3) concordance of results among mesocosms, laboratory studies, and field observations will provide important weight of evidence that can be used to refine the conceptual model and provide a robust predictive model for use by risk managers.

The results of initial applications of the test scheme will be used to refine the initial risk hypotheses and conceptual model. The overall goal is to use repeated iterations of this test scheme to produce a robust and flexible conceptual model that can be used to predict ecosystem responses to multiple stressors in a wide variety of situations.

A weight of evidence would be assembled using the results of the experimental mesocosm tests, supported by experimental laboratory data and observational field data. Weight of evidence is the process by which laboratory, mesocosm, and field effects endpoints are evaluated with respect to their concordance with the multiple-stressor risk hypotheses. The proposed method follows a three-step process:

Step 1: Establishing causality
Step 2: Defining relative risk
Step 3: Summing up the weight of evidence in a decision-making framework.

Establishing causality requires that the triad of field, laboratory, mesocosm, and field data produced by the General Test Scheme must be examined for cause/effect relationships between the stressors and endpoints. Without strong evidence for causality, the multiple-stressor hypotheses cannot be successfully tested. Distinguishing among the cumulative effects of multiple stressors (which sometimes have confounding effects) requires establishing a strong

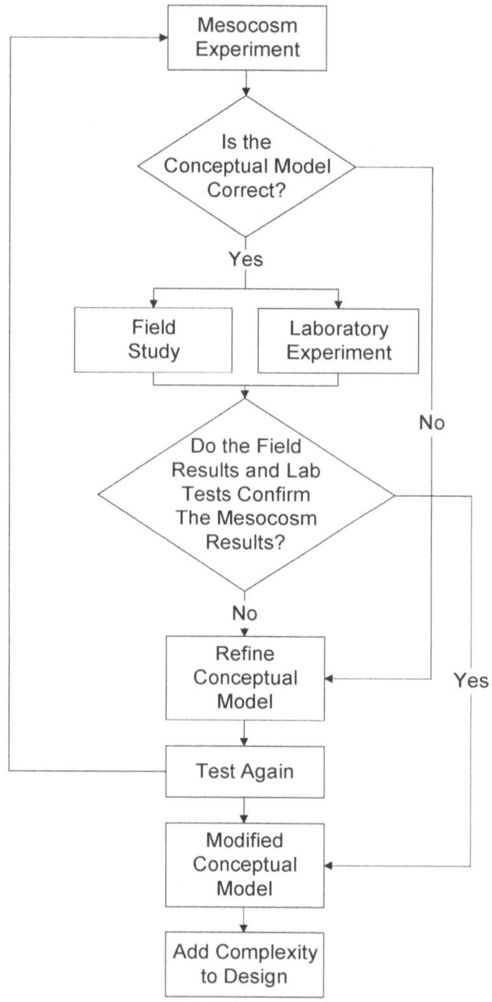

Figure ES-3. General Testing Scheme.

causal link to each stressor. This requires evaluation using a formal set of criteria.

Defining relative risk requires that each endpoint be evaluated against some benchmark of "relative risk." The definition of relative risk should be as clear and explicit as possible. The definition will always involve the use of professional judgment; however, evidence in support of the definition from the scientific literature should be provided.

Assembling the weight of evidence amounts to determining what overall risk statement is best supported by the individual lines of evidence. If the benchmarks for relative risk have been carefully and clearly defined, the production of a summary weight-of-evidence table should be quite straightforward. The "art" involved in summary tables is the development of logical statements in support of the "overall risk ranking" derived from the summation of relative risk scores for all endpoints. Examples presented in this study show how data generated by the General Test Scheme for the conceptual model can be interpreted within a risk-based weight-of-evidence framework. Actual data will be more complex, and will involve a longer list of endpoints. However, if the basic three-step process is carefully followed and documented, the interpretation of the weight of evidence from laboratory, mesocosm, and field data should be clear and defensible.

ES.5 Details of the Study Design

Although the details of the study design used to test the conceptual model will be left to future multiple-stressor research, this report provides guidance regarding the components of the study design, as briefly summarized below.

The study team recommends that testing progress through the following series of stressor combinations: toxicant plus toxicant, then subsidy plus subsidy, and finally toxicant plus subsidy. This recommended order of testing avoids ambiguous situations by choosing combinations that are firmly in one camp or the other. Because the subsidy region of the concentration curve is small relative to the toxicant region (and for some stressors, may not exist), it is best to begin with combinations of toxicants. If initial toxicant tests are successful, the more difficult studies combining subsidies could be undertaken. Only when these two options have been adequately explored should we attempt combinations of a subsidy and a toxicant.

There are many mesocosm designs available with a range of advantages and disadvantages. The ultimate choice of mesocosm design depends on an evaluation of the strengths and weaknesses of each candidate system versus criteria such as cost, practicality, portability, flexibility, ability to incorporate several food chain components, and the ability to manipulate important environmental variables such as flow, light, and temperature.

The specific field locations for the mesocosm and confirmatory field studies will be determined at a later stage. To date, it is agreed that study sites will consist of lotic systems that are of similar stream order and of comparable size/flow, and that are representative of those typically receiving point discharges in the region. Candidate streams and rivers should be neither too small nor too large; larger waters are likely to contain a multitude of multiple stressors, which would add too many confounding variables. The streams and rivers should have access to riffle communities and moderate-to-low flow variability.

Logistical criteria for site selection would include an existing water supply system, a nearby wastewater treatment plant willing to accept mesocosm wastewater for treatment, ready site access for at least two seasons, potential for long-term experimentation, no undue complexities regarding obtaining permits and good security. Collaborator criteria would include proximity to laboratory facilities, concurrent research activity, the existence of historical data, and willing participants.

The sites should only receive a few point source stressors in the study area ("reach" of the river). To keep the number of variables to a manageable level, and avoid confounding effects as much as possible, the reach of the river to be tested should be fairly uniform with few point source contributions. Other environmental, logistic, and collaborator criteria will be key in the site-selection process.

Candidate sites suggested by the study team and the PSC are located on the Willamette River, Oregon; the Little Miami River, Ohio; and the Wapiti River, Alberta, Canada.

Stressor selection will use criteria based on stressor importance, the ability to manipulate and measure, and sociopolitical considerations. Stressor importance will be judged according to how common the stressor is in North American rivers, the persistence or continuous nature of loading of the stressor, the availability of background data, proven toxic potential, and the potential to interact with other stressors. Sociopolitical relevance will be defined by the degree of societal concern about a particular stressor, combined with the regional presence of the stressor.

The study team recommends that the selected stressors be a combination of physical and chemical stressors commonly present in the real world. Some stressors to be considered for use in experiments are:

Chemical Stressors

- Metals such as copper, nickel, or zinc.
- Nutrients such as phosphorus or nitrogen.
- Dissolved organic carbon.
- Surfactants such as alkyl sulphate.
- Herbicides such as atrazine, 2,4-dichlorophenoxyacetic acid (2,4-D) , or glyphosate.

Physical Stressors

- Fine sediment (i.e., sediment that causes increased embeddedness).
- Elevated temperature.
- Flow variability.
- Ultraviolet radiation.

The endpoints to be measured in the multiple-stressor experiments derive from the conceptual model being tested and the structure of the artificial ecosystems being assembled in the mesocosms. For simplicity's sake, the mesocosms will have three trophic levels: primary producers (benthic algae), primary consumers (benthic macroinvertebrates), and decomposers (benthic microorganisms). Endpoints should be selected from each of these trophic levels.

Fish will be excluded from the mesocosms because of the difficulties in maintaining sufficient numbers and diversity to observe stress responses. However, secondary consumers may still be represented in the mesocosms by predatory invertebrates such as leaches and stoneflies.

The mesocosm experiments must test the hypothesis of additivity between multiple stressors applied to flowing-water ecosystems. An analysis of variance (ANOVA) approach is recommended because it is more robust for testing the effects of stressor combinations.

The number of treatments in the ANOVA design is a compromise between the amount of information gained (i.e., the confidence in the conclusions) and the budget. In the first phase of the experimental work, selected stressors will be applied individually to mesocosm communities to document the nature and degree of response to different stressor levels, one stressor at a time.

For these experiments, the stressor levels should be chosen to span the natural range of levels likely to occur.

Once the single-stressor experiments have established the shape of the response curves for each selected stressor under mesocosm conditions, appropriate combinations can be selected for testing in combination. Levels of each stressor that are clearly in the stimulatory (subsidy) or inhibitory (toxicant) range for each stressor should be selected. Stressor levels that are close to a cusp between stimulation and inhibition or that are otherwise at the extremes of the natural distribution should be avoided, as they are more likely to produce ambiguous results.

The experiments to test effects of multiple stressors would use a full two-way ANOVA in which every level of Stressor A is tested at every level of Stressor B, as illustrated in Table ES-1.

Table ES-1. Full Two-Way Analysis of Variance.

		Stressor A		
Stressor B		**Low**	**Medium**	**High**
	Low	1: Low/Low	4: Medium/Low	7: High/Low
	Medium	2: Low/Medium	5: Medium/Medium	8: High/Medium
	High	3: Low/High	6: Medium/High	9: High/High

The design permits tests of the main effects of Stressor A and Stressor B, and the interaction between them. The last test is pivotal. If the hypothesis of additivity is correct, the *F* value (amount of difference between results) for the interaction effect will be nonsignificant. Any departure from strict additivity, either synergistic (more than additive) or competitive (less than additive), will lead to a significant *F* for the interaction effect in the ANOVA.

ES.6 Cost of Implementing the Study Design

The estimated total cost involved is approximately $600,000 USD, to be divided over a 2-year period. This is a rough estimate because many decisive factors are yet to be decided: the study location and final design, type of equipment used, site logistics such as ease of access, convenient waste disposal facilities and proximity to analytical laboratories, the type of experiments conducted, and the choice of mesocosm apparatus used.

CHAPTER 1.0

INTRODUCTION

1.1 Background and Summary of Recent Multiple-stressor Initiatives

The subject of multiple-stressor effects in aquatic ecosystems has been of interest to scientists and environmental managers for some time (e.g., the review papers in *Limnology and Oceanography*; Breitburg et al., 1999). There is a broad consensus on the need for a better understanding of multiple-stressor effects; however, there has been limited progress in developing and testing a general set of theories for multiple stressors.

Multiple-stressor effects was the topic of a 1997 Society of Environmental Toxicology and Chemistry (SETAC) Pellston workshop (Foran and Ferenc, 1999). The workshop developed a framework to assess, compare, and manage the effects of multiple stressors on biological systems. The specific objectives were to assess the state of knowledge on incorporating multiple stressors in ecological risk and impact assessment and to develop a framework and specific methods and models for the qualitative and quantitative evaluation of multiple stressors. Another key objective was to determine how multiple stressors might be identified and assessed through the structure and function of ecological systems.

The major difficulties encountered during the SETAC workshop included:

- Different terminology used by the scientists involved in risk assessment for multiple stressors and in cumulative impact assessments
- The timeframe used by each type of assessment (the first assessment as "retrospective" and the final assessment as "prospective")
- The approach to the process: risk assessments perceived as "stressor-driven," while impact assessments were driven by effects analyses
- Multiple stressors were usually conceived as a single or perhaps multiple chemicals only

The workshop had several agreed-upon outcomes:

- Risk assessment for multiple stressors and cumulative impacts assessments can be either retrospective or prospective (with the law or statute requirements dictating the timeframe to be used).
- A variety of stressors shape ecosystems and must be considered in both types of assessment.
- Multiple stressors are to be perceived as any radiological, chemical, biological, or physical alteration to the ecosystem, of both anthropogenic and natural origin.

Understanding that there is rarely a single stressor involved in observed damage to an ecosystem constituted a major shift to the approach to the investigation of multiple-stressor effects. This approach supports previous studies in this field (Karr et al., 1986; Yoder and Rankin, 1995), who have identified the most important factors affecting an ecosystem's function and structure (Figure 1-1).

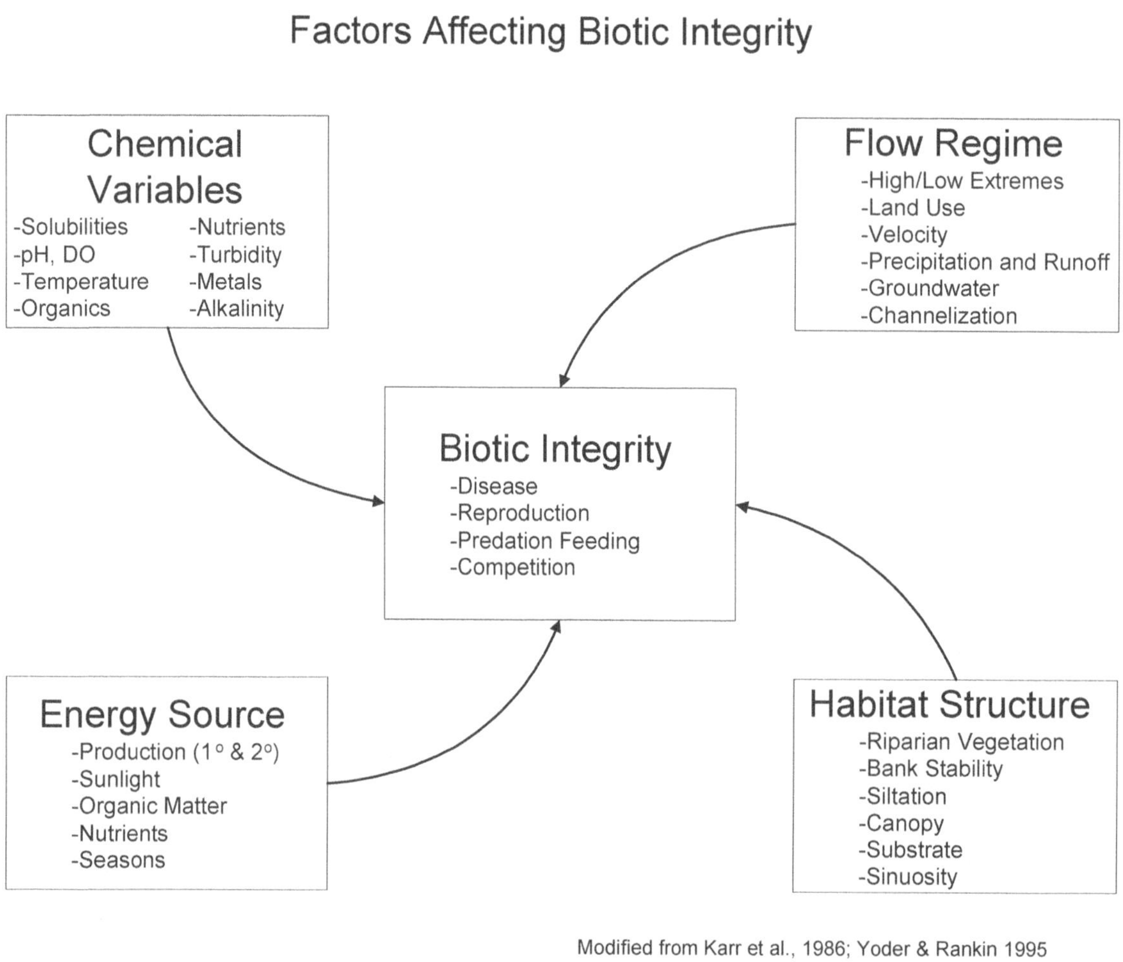

Figure 1-1. Factors Affecting Biotic Integrity.

The SETAC workshop was the first of several recent efforts to develop methods to assess the effects of multiple stressors in ecosystems. It was recommended that a second workshop "develop/identify an approach to determine a common metric by which to measure effects of different stressors in ecological systems" (Foran and Ferenc, 1999). At the time, no available conceptual models addressed the quantitative assessment of multiple stressors.

Following the Pellston workshop, Foran and Ferenc commissioned the development of five chapters about the methods used in assessments of multiple stressors. That work (Foran and Ferenc, 2000) discussed methods for risk estimation. The book contained a number of useful tools, including modeling and statistical approaches and effect-based assessment methods. However, the book did not present a unified theory of multiple-stressor effects in ecosystems; therefore, it presents the tools but contains no hypotheses to test.

Another response to the need for tools for multiple stressors was the publication of U.S. Environmental Protection Agency's Stressor Identification (SI) Guidance Document (U.S. EPA, 2000). The document provides the user with valuable tools to screen for a stressor(s) that may have caused an observed (or measured) biological impairment in a given waterbody. This tool can be very valuable for decision-makers for particular uses because of the rigorous procedures involved, as well as the following aspects: as an iterative and scientific process, it can produce reliable and quite accurate results in the identification of the stressor(s) most likely responsible for the observed biological impairment; its nature allows for building a thorough and valuable inventory of stressor(s) related to the particular damage; and it is based on weight of evidence.

Because it is based on observational data, this tool can only provide the user(s) with a snapshot (retrospective) view of the observed biological damage and identified damage-causing stressor(s). Other limitations are:

- Based on observational data, the tool draws its conclusions on associations and does not lend itself to predictions.
- The scope of its application is very site-specific (i.e., its outputs are pertinent to investigated sites and to observed damages).
- This type of tool may be time consuming and involve repetitive work that has to be redone for every single site because much of the valuable information collected locally is actually unusable for other ecosystems.
- Observational data are not useful for accurately pinpointing the causative stressor(s) when multiple stressors are present because the nature of their interactions is not yet well established.
- The assumption that multiple stressors interact in an additive manner is an oversimplification of the problem, as many observational studies have determined. (Harwell and Gentile, 2000; Fairbrother and Bennett, 2000; Rosenberg and Resh 1993)

One example of the SI tool's limitations is when a new point source discharge containing "alien" (not present before) stressors is introduced to a particular site, or when a known source discharges new stressor(s). There is a strong possibility that the completed "inventory" becomes invalid or seriously outdated, because this tool cannot predict the type and/or magnitude of the impact(s) caused by the newly introduced stressors. This limitation would not apply to ecosystems that are already severely affected, because their biological make-up would be already very simplified.

1.2 Experimental and Field Approaches to Multiple-stressor Research

There are a number of experimental and field observational methods used in stream ecology that can be applied to the study of multiple stressors. Experimental methods include semi-natural contained systems (microcosms and mesocosms). Microcosms and mesocosms can have a variety of sizes and configurations and may or not incorporate flowing waters. They are used as a bridge between laboratory and field experiments. Mesocosms provide the most potential for experimental investigation of multiple-stressor effects on communities and ecosystems. A detailed review of mesocosm design is presented in the accompanying literature review report (Swanson et al., 2004) and briefly in Chapter 4.0 of this report (Experimental

Design). The following text provides an overview of mesocosm approaches, as well as field approaches.

The value of mesocosms as predictive bioassessment tools has been discussed in a variety of forums (Culp et al., 1996, 2000b; Clements and Kiffney, 1996; Barbour et al., 1996; Kline et al.,1996; Dorn et al., 1996; Hill et al., 1994; Graney et al., 1994; Baker et al., 1994; Lamberti and Steinman, 1993). These tools allow realistic multispecies tests to be performed under replicated conditions. Although generally used to measure population and community level responses, the tools have also been used to measure ecosystem-level structural and functional endpoints such as production, biomass, respiration, nutrient dynamics, resistance, and resilience (Beyers and Odum, 1993; Lehtinen et al., 1993; Webber et al., 1992; Fairchild et al., 1992; Perry and Troelstrup, 1988; Boyle et al., 1996). Reviews of methods in stream ecology (including stream mesocosms) include Beyers and Odum (1993). Mesocosms and microcosms have definite utility for testing theories of multiple-stressor effects (Lamberti and Steinman, 1993; Hauer and Lamberti, 1996).

Beyers and Odum (1993) critically reviewed and evaluated various published studies using experimental stream microcosms under varied conditions:

- Study of relationships among several stressors (e.g., light, circulation rate, nutrients and organic matter);
- Affects on stream ecosystems by the addition of multiple stressors such as chemical discharges (e.g., chlorine, chromium, copper, sucrose);
- Studies on the degradation rates of toxic substances in aquatic ecosystems (e.g., P-cresol); and
- Study of long-term assimilation capacity of streams subjected to multiple stressors.

Lamberti and Steinman (1993) edited a thorough evaluation of the current status of artificial stream research and especially its contribution to the understanding of natural systems. They reviewed a large number of published studies using various types of artificial streams used for different purposes (e.g., for grazing studies, fish ecology, ecotoxicology) and identified the advantages and limitations of each approach. They arrived at the following conclusions:

- There is no single best design for artificial streams, and each is contingent upon the questions being asked.
- Artificial streams are particularly well-suited to research geared to mechanistic understanding of lotic systems.
- The generation of testable hypothesis, which can be validated in natural stream ecosystems, is an extremely useful application of research in artificial streams.

Hauer and Lamberti (1996) edited an analysis of the various methods used in stream ecology. Most of these methods include stressors and receptors used to investigate multiple stressors: physical processes (e.g., stream flow analysis, temperature, light, oxygen); material storage and transport (e.g., siltation, nutrient loadings); stream biota (e.g., study of receptors: heterotrophic microorganisms, algae, benthic macroinvertebrates, fish communities); community interactions (e.g., plant-herbivore interactions, trophic relations of macroinvertebrates and fish, habitat use and stream food webs); and ecosystem processes (e.g., primary productivity and community respiration, secondary production of macroinvertebrates, effects of nutrient enrichment on periphyton). Their book is useful to anyone involved in stream ecology research because it contains a number of general study designs and specific applications.

Some extensive large-scale outdoor experimental stream studies were conducted over time to study the effects on stream biota caused by point source discharge from a complex pulp and paper mill effluent (Hall et al., 1991; Haley et al., 1995). Studies included monitoring of potential multiple stressors such as groups of organic chemicals, biological oxygen demand (BOD), color, turbidity, dissolved oxygen (DO), total suspended solids, and nutrients. The receptors studied were fish, macroinvertebrates, and periphyton. The researchers monitored both stressors and receptors for effects from the original bleached-Kraft Mill effluent and to effluent produced following conversion to a chlorine dioxide bleaching process.

These studies contributed to a better understanding of the effect of color and total suspended solids on aquatic biota and demonstrate the utility of large-scale, long-term artificial stream studies.

Field studies of the multiple-stressor effects of complex effluents include several studies of large western Canadian rivers that receive pulp mill effluents [as well as other point and nonpoint source (NPS) discharges] (Swanson et al., 1994; Lowell et al., 1995a,b, 1996, 2000; Culp et al., 2000a). Several regional- or watershed-scale studies are underway in the United States. For example, so-called scaled assessments of risks to the Columbia River system are currently being conducted, using hierarchy theory to determine efficient and cost-effective scales for assessing and monitoring the condition and sustainability of aquatic systems within the Columbia Basin (Goodman et al., 1998). A statewide characterization of the effects of multiple stressors in Ohio used Geographical Information System tools and multivariate regressions to screen for relationships between habitat, water chemistry, and biotic integrity (Dyer et al., 1998). U.S. EPA, U.S. Fish and Wildlife, and state agencies are conducting studies of watershed-scale or regional-scale relationships among stream chemistry, physical habitat, land use, and biologic condition in the Mid-Atlantic region and the Midwest region of the United States (e.g., Smith and Voshell, 1998; Jones et al., 1998; Stoddard et al., 1998; White et al., 1998).

Current large-scale, long-term field research programs by the National Council on Air and Stream Improvement (NCASI) focus on watershed-level assessment of effects of pulp and paper mill effluent discharges. The studies take into account other complex stressors at the basin level (NPS agricultural discharges, urban development, habitat degradation, thermal discharges, and several point sources of municipal wastewater discharges). These observational studies (which include some mesocosm studies as well) produced large databases (Hall, 2001) that can be used to help validate the proposed conceptual model presented in the report in later stages.

These large-scale studies usually focus on the search for pattern in the response indicators (e.g., trends in species richness) as well as correlations between response patterns or trends and land use variables or particular point and nonpoint sources. These studies generated a number of useful insights into landscape-scale trends. However, some reviewers expressed concern over the approaches taken in some of these large-scale assessments, pointing out that the derived indices often used reflect regulatory management practices rather than endpoints representative of inherent physical, chemical, or biological health (White-Hull et al., 1998).

1.3 WERF Workshop on Multiple Stressors

A 1997 WERF-sponsored workshop, "Critical Research Needs for Understanding Ecosystem Risks from Multiple Stressors" examined the need for a framework to address ecosystem-level, risk-based approaches to assess and manage multiple stressors in aquatic ecosystems (Gerritsen et al., 1998). The workshop report (97-IRM-3) indicated that the decision framework should emphasize the following:

- An integrated decision-making process that involves policymakers, resource managers, regulators, scientists, and other stakeholders to make effective ecosystem management decisions.
- Developing qualitative and quantitative measures of value for ecological goods and services involves: a balance between factors that are ecologically relevant and those that are of value to society, a determination of when to apply management to ecological changes, and a determination of what effects will be ecologically significant.
- The importance of partitioning and ranking the effects of multiple stressors.
- Integrating natural and socioeconomic sciences in the decision-making process.

The report also emphasized the use of both observational and experimental research to predict or detect patterns caused by multiple stressors. The highest priority ecosystem research needs identified in the workshop report were to review modeling approaches; review and synthesize existing information; conduct long-term studies of stressed ecosystems; develop cumulative impact assessment methodologies; and conduct descriptive and experimental studies of stressed ecosystems.

1.4 Other WERF Multiple-Stressor Projects

Two recent WERF projects use the observational approach in the search for patterns of response to multiple stressors. Brown and Caldwell (Paulson et al., 2001) investigated the ability to discriminate habitat versus chemical limitations using novel data mining techniques. They also provided a comprehensive review of bioassessment methods and the use of metrics to describe the status or "health" of receiving streams. Gerritsen and Jessup (2004) investigated the response of stream invertebrates to multiple stressors using data sets from Florida, Wyoming, and West Virginia. These studies resulted in the development of useful tools in multiple-stressor field research and data analysis, but could not extract the underlying mechanisms behind the interaction of stressors on receptors in a multiple-stressor situation.

Brown and Caldwell (Paulson et al., 2001) point out that although metrics adequately characterize the differences in aquatic populations, they cannot define what specific chemical or habitat limitations affect the aquatic community. They also show the challenges inherent in identifying and agreeing on reference streams. Pristine streams with comparable physical conditions and habitat are hard to find. They emphasize the importance of distinguishing between the effects of habitat degradation and chemical stressors because of the current efforts by the U.S. EPA and state agencies to implement Total Maximum Daily Loads (TMDLs) for streams. It will be very important to determine how best to spend money on environmental improvements in reductions of chemical loads or in habitat improvements. Brown and Caldwell (Paulson et al., 2001) suggest that the traditional reliance on inferential statistics limits our ability to determine complex cause-effect relationships among multiple stressors and the receiving environment. They provide examples of data-mining tools such as Principal Components Analysis, All Possible Regressions, and Chi-Square Automatic Interaction Detection to identify key variables in large environmental data sets.

Brown and Caldwell (Paulson et al., 2001) demonstrated the utility of data mining in the search for patterns of response to multiple stressors using four data sets. In the Santa Ana River (California), they found that a combination of habitat characteristics, ammonia, and chlorine— not metals— caused biological impairments in the stream. Thus, the wrong mitigative measures may have been applied had not additional analyses been carried out. They also found that the data mining techniques worked better on data sets that were not highly diverse; investigation of the Cuyahoga River (Ohio) data set showed how a wide range of independent variables can make it difficult to detect significant interactions between dependent and independent variables. Nevertheless, key variables were identified that included total number of macroinvertebrates versus cover and water hardness, and relative weight of fish versus DO and hardness.

The importance of matching habitat data was illustrated in the South Platte River (Colorado). The lack of habitat data made it impossible to determine if there was an affect to biota from habitat changes over time. However, the project team was able to develop a model of the Shannon Weaver diversity score versus orthophosphate and flow. The Fountain Creek (Colorado) dataset was used to validate a User Guidance Document (Paulson et al., 2001) on the data mining techniques. The model developed from Fountain Creek described invertebrate taxa richness as a function of stream channel embeddedness and orthophosphate concentrations. Graphical examination of the data revealed a linear decrease in taxa richness with increasing embeddedness, whereas orthophosphate had a marginal effect on taxa richness until it reached high levels.

Gerritsen and Jessup (2004) identified and developed predictive, diagnostic relationships that identified causes of impairment. Results of the research would form part of the evidence in SI using U.S. EPA's guidance (U.S. EPA, 2000). The project team used three large databases from Florida, Wyoming and West Virginia in their search for indicators of multiple stressors. Habitat quality was the single most important in-stream stressor. Metrics that responded to the presence of single stressors always showed an association with habitat. Association of the metrics with multiple stressors was one of degree, the greater the number of stressors, the greater the degradation of the median metric values. The study team was unable to distinguish unique or diagnostic responses of metrics to individual stressor types.

Although these studies provide tools to help with observational studies and our understanding of the effects of stressor interaction, the mechanisms of interaction in multiple-stressor situations are yet to be determined.

1.5 Literature Review Summary

A literature review was conducted in support of this study design document to determine the state-of-the-science pertaining to multiple-stressor research (Swanson et al., 2004). The literature review focuses on experimental studies.

A rigorous, theoretical basis for the prediction of multiple-stressor effects does not exist. Despite the wealth of observational data, the existence of several useful tools for interpretation of cause/effect relationships (including formal SI methods) and the studies reviewed in the literature review report, there are no tools that allow a confident, *a priori,* prediction of ecosystem response to multiple stressors. Retrospective studies can use weight-of-evidence approaches and the standard rules for establishing cause/effect relationships to piece together a plausible story regarding the stressors most likely to have caused the observed responses. However, these retrospective studies are time-consuming, data-intensive, and result in analyses that cannot be readily extrapolated to other cases. Furthermore, uncertainty associated with these retrospective studies still centers around model error (i.e., our imperfect understanding of how multiple stressors can be expected to interact and how ecosystems can be expected to respond). Instead, we default to a standard assumption of additivity (in the case of chemical stressors) or to a case-by-case, professional judgment process.

The ultimate goal for the assessment of risks from multiple stressors is to have an understanding of multiple-stressor effects that is sufficiently robust to support decision making. In other words, we aspire to combine observational and experimental evidence for pattern and process responses to multiple stressors in a manner that allows confident analysis and prediction.

A common finding throughout the multiple-stressor literature is that the interactive effects of these stressors are not addressed, largely because it is difficult to apportion the effects observed to the variety of stressors present. Studies that use the weight-of-evidence case for cause and effect relationships among stressors and receptors make an attempt to address these interactive effects.

There is considerable difficulty in quantifying the effects of multiple stressors that interact in a nonlinear manner. With the benefit of extensive research on specific interactions, we can produce reasonable predictions of cumulative stress (e.g., from many years of study of aluminum plus pH). It is a greater challenge to calculate the cumulative stress from two stressors whose interaction has not been thoroughly studied.

The current literature provides, at best, a series of site-specific glimpses of the response of ecological communities and ecosystems to multiple stressors. The definition of the problem provided by authors is usually very focused on a particular site or a particular combination of stressors. There is seldom, if ever, any reference to a more generalized model of multiple-stressor effects apart from the discussion of the expectations regarding additivity versus synergism. Many articles that purport to be multiple-stressor studies do not go beyond an inventory of the various stresses and upsets affecting the ecosystem, without attempting to assess the interactions among them. Even the regional risk-ranking model of risk assessment championed by Landis and colleagues (1998, 2000) merely add up all the stresses affecting different areas within the region, and do not attempt to compute their interactive effects.

The exceptions to these studies are papers that define the problem in terms of building a weight-of-evidence case for cause/effect relationships among stressors and receptors, or papers that refer to a suggested general response pattern to stressors (e.g., Munkittrick and Dixon, 1989; Gibbons and Munkittrick, 1994). However, even these papers seldom place the definition of the problem in a wider framework provided by a general model such as Odum's 1985 model of ecosystem stress.

One notable paper by Lewis (1999) presents three testable hypotheses concerning general responses of ecosystems to stress. They are: 1) a given biotic compartment in an incrementally stressed ecosystem will show suppression of biodiversity first and suppression of biomass subsequently (similar to the idea expressed by Schindler in 1987); 2) suppression of ecosystem function is primarily caused by reduced biomass, rather than loss of diversity itself; and 3) suppression of biomass differs among biotic compartments according to a predictable series of thresholds (i.e., multicellular organisms are more sensitive than unicellular autotrophs algae, which are more sensitive than microbial decomposers). The third hypothesis implies a predictable sequence of ecosystem processes that will be lost as the severity of stress increases. Lewis's work is seminal in that it goes beyond simply listing the responses of a particular system to a particular set of stressors, but instead provides a general, testable model of ecosystem behavior. Unfortunately, work to test these hypotheses has only just begun, although Lewis (1999) suggests that preliminary studies on streams suffering acid mine drainage are consistent with his hypotheses.

Because of the lack of a broader context to multiple-stressor research, the current literature does not provide a readily accessible basis for the construction of generalized conceptual models of multiple-stressor effects on aquatic ecosystems. However, a few tentative generalizations may be possible based on research to date and extrapolation from broader ecological theory. These generalizations are:

- The organizational level that is being studied will determine the type and additivity of effects.
- Generally, it appears that stresses of most kinds have a simple additive effect on individual organisms, but the response to stress is nonlinear.
- We can anticipate that populations and communities will respond to stress in a predictable, linear fashion (above a lower threshold) because increasingly less-sensitive individuals or species will be removed as the intensity of the stressor(s) increases.
- The effect of multiple stressors is probably more often additive than synergistic or antagonistic at the population level.
- According to Odum (1985), stressed ecosystems maintain processes at the expense of species; consequently stressors may have sharp thresholds of effect, below which the effect is minor or even undetectable (still enough species left to do things). Above this level the effect is abruptly pronounced (no longer enough species or individuals to support essential processes).
- If the stressor is stimulatory of growth or production, like heat or nutrients, then a subsidy effect may be observed at low levels (Odum, 1985). The subsidy side of the equation in

Odum's model has been given less consideration than the stress side. Nevertheless, scattered observations in the literature broadly support Odum's conception.

- Even toxic chemicals may act as subsidies to the ecosystem if the concentration is low enough.
- The nature of interactions between stressors and subsidies will determine the overall ecosystem response to multiple stressors.
- It is important to distinguish the interactions between the stressors from the interactions between stressor effects. Much of the confusion about interactions between broadly different stressors arises from a failure to make this distinction.
- Complex and apparently daunting stressors (i.e., climate change, invasive species, overharvesting, and flow regulation) can perhaps be rendered tractable by carefully and painstakingly considering their effects on ecosystems. The risk assessment can concentrate on precisely explicated changes to the system.

There are no examples of explicitly identified assessment endpoints as defined within the risk assessment framework for experimental studies of multiple stressors in the literature. Conceptual models are also missing, except for the occasional mention of the expectation of additive versus synergistic effects.

Assessment endpoints must be measurable ecosystem characteristics that adequately represent management goals (U.S. EPA, 1998). For example, if the management goal is to reduce the risk to striped bass populations from multiple-stressor interactions to acceptable levels, then the assessment endpoint must be sensitive to the identified stressors, yet respond in different ways to the different stressors. Striped bass populations may be adversely affected at several life stages, in different habitats, through different ways and by different stressors. A useful assessment endpoint would stipulate the attributes of striped bass populations most likely to respond to the particular combination of stressors under investigation. This assessment endpoint would be accompanied by a set of measures that encompass the range of responses of the various life stages to the different stressors.

Conceptual models consist of two main parts, a set of risk hypotheses that describe predicted relationships among stressor, exposure and assessment endpoint response, and a diagram that illustrates the relationships in the risk hypotheses (U.S. EPA, 1998). Risk hypotheses are specific assumptions about potential risk to assessment endpoints and may be based on theory and logic, empirical data, mathematical models or probabilistic models (U.S. EPA, 1998). There was a notable lack of risk hypotheses meeting these requirements in the literature. Conceptual model diagrams are a visual representation of these risk hypotheses. There was also a notable lack of conceptual model diagrams in the literature.

The lack of multiple-stressor conceptual models is a primary source of uncertainty in the assessment of multiple-stressor risks. Because of this lack, the risk characterization of multiple stressors may misinterpret the actual risks.

"Multiple stressors are almost always present, and can be the source of confusion, especially when conceptual models focus strictly on single stressors" (U.S. EPA, 1998). This general, and rather obvious, statement by the U.S. EPA is, in practice, seldom or never addressed explicitly in the literature except perhaps by the use of a general assumption of additivity.

Therefore, any contribution to the development of general conceptual models for multiple-stressor situations will be very valuable and is the focus of this report.

Field observational studies are not sufficient to provide a sound theoretical basis for understanding multiple-stressor effects in aquatic ecosystems. Although observational studies have been useful in identifying the general attributes of systems exposed to multiple stressors, the actual role of particular stressors versus the role of the mixture of stressors is difficult to deduce from these data. Furthermore, the ecological processes leading to the observed responses can only be speculated upon.

The generation of testable hypotheses is probably the most useful aspect of observational research in multiple stressors.

An experimental approach is required to test hypotheses arising from conceptual models of multiple-stressor effects. Experimental results can be used in combination with observational field studies in an iterative manner to produce increasingly robust conceptual models.

The study design presented in this report includes provision for complementary field studies; however, its emphasis is on the experimental approach because of the demonstrated lack of a sound experimental basis for multiple-stressor theories.

CHAPTER 2.0

STUDY DESIGN SCOPE

2.1 Goals and Objectives

The goal of this project is to provide a study design based on good science that helps establish a general, conceptual approach to multiple stressors. The objectives are:

- Develop a flexible and scientifically defensible conceptual model of the environmental effects of multiple stressors on river and stream ecosystems.
- Develop a study design to test the conceptual model of multiple-stressor effects.
- Document the process used to arrive at the conceptual model.

2.2 Project Duration

The project consisted of conducting reviews of the scientific literature (e.g., specific case studies, theoretical papers, etc.), consultation among team members to develop the conceptual framework, meetings with the Project Subcommittee (PSC), and report writing. This process allowed for underlying theories to be developed and for formulation of the conceptual model used in the ecosystem design. The proposed experimental portion of this work, which will be used for testing and validation of the conceptual model, is designed to last for approximately two years. It is important to note that the testing of a conceptual model is an iterative process, so only a few variables are introduced in the early stages of the study. Subsequent iterations of the study will examine additional combinations of stressors to produce further refinements of the conceptual model and broaden its applicability.

2.3 Types of Investigations Encompassed by the Study Design

The study design allows for the integration of laboratory testing, mesocosm studies, and confirmatory field investigations. The study design has an experimental emphasis because the literature review showed that there is a significant lack of experimental studies compared with field observational studies. Experimental studies are required to meet the goal of this project (i.e., to test risk hypotheses to produce a general, conceptual approach to multiple stressors). Laboratory experiments are limited in their ability to adequately represent multiple-stressor situations. Therefore, the study design focuses on outdoor mesocosms, which represent a compromise between the ability to control variables and establish cause/effect relationships and the need for greater environmental realism.

Laboratory studies will be used in a supporting role in this investigation. For example, they can be used to determine specific stressor threshold levels. They can also be used to help confirm cause/effect relationships between a particular combination of stressors and a particular endpoint.

Field studies will be used to help validate the results of mesocosm experiments. For example, if the mesocosm experiments predict that the combination of stressors present at the study site would produce a certain reduction in periphyton biomass and a shift to tolerant species, field data can be collected to confirm this prediction.

The integration of these three tools (laboratory, mesocosm, and field studies) in a weight-of-evidence approach is described in detail in Section 4.2.

Specific field locations for the mesocosm and confirmatory field studies will be agreed upon at a later stage (refer to Chapter 4.0). To date, it is agreed that study sites will consist of lotic systems that are of comparable size/flow, and that are representative of those typically receiving point discharges in the region. Stream order is not a determining criterion, as it is basin-specific and not closely related to size or flows. Candidate rivers should not be too small, but rather, be representative of a region, but also not too large as they likely contain a multitude of multiple stressors, which would add far too many confounding variables. The sites should only receive a few point stressors in the study area (reach of the river) and the rationale behind limiting the number of effluent discharges is to keep the number of variables to a manageable level, thereby avoiding confounding effects as much as possible. In other words, the reach of the river to be tested should be fairly uniform with few point source contributions. Other environmental, logistic, and collaborator criteria will be key in the site selection process.

2.4 Outline of the Process Used for Design Development

The development of the proposed conceptual model, associated risk hypotheses and study design approach was a creative and iterative process. The literature review provided an excellent starting point. The literature review confirmed the presence of a fundamental gap in our understanding of multiple-stressor effects in ecosystems (i.e., there were no published "unified theories" of multiple-stressor effects). The reliance on the assumption of additivity, the dominance of field-based observational studies, the lack of experimental studies on relevant mixtures of stressors, and the lack of predictive capability all contributed to the commitment of the study team to providing a fundamental, theoretical starting point for subsequent research.

There were several important features of the process used to develop the conceptual model and study design. These features were:

- An open approach to new ideas where every suggestion was considered.
- Early agreement among team members that, at this stage in our understanding, the best approach is to develop a general concept and then test that concept.
- Early agreement among team members on a strong experimental emphasis that integrates laboratory, mesocosm, and field studies.
- A technique of working backward from knowledge gained through observational data and forward from the conceptual model perspective.
- Openness to the use of older, solid scientific advances in this area (including the fundamental ecological theories developed in the 1960s and 1970s).

The members of the team represented several relevant scientific areas:

- Stream ecology and mesocosm research
- Stream ecotoxicology and biostatistics
- Large river ecotoxicology and mesocosm research
- Field ecotoxicology and risk assessment

The combination of having key specialists and an open forum for discussion of ideas was very successful in the development of the conceptual model presented in this study design.

CHAPTER 3.0

CONCEPTUAL MODEL FOR ECOSYSTEM RESPONSES TO MULTIPLE STRESSORS

3.1 Review of Relevant Terminology

The proposed conceptual model uses a risk assessment framework and terminology that is briefly reviewed here.

Risk assessment is a standardized approach to the evaluation of the probability and consequences of a change in ecosystem structure or function. A systematic approach to ecological risk assessment has been described in "Guidelines for Ecological Risk Assessment" (US EPA, 1998). This approach includes three primary phases: problem formulation, analysis, and risk characterization and is the approach adopted as the overall risk framework for this report.

Problem formulation is a process for generating and evaluating preliminary hypotheses about why ecological effects have occurred, or may occur, from anthropogenic causes. In the problem formulation, the objectives of the risk assessment are refined, and the problem is evaluated so that a plan for data analysis and risk characterization can be developed (U.S. EPA, 1998). Thus, it is the foundation for the entire ecological risk assessment.

There are three main products from the problem formulation:

- Assessment endpoints that adequately reflect the ecosystem and management goals.
- Conceptual models that describe key relationships between stressors and assessment endpoints.
- A plan for analysis of these relationships (U.S. EPA, 1998).

The focus of this report is on problem formulation, with emphasis on the production of a conceptual model with associated risk hypotheses, as well as on suggested stressors, receptors, assessment endpoints and measures for use in testing the risk hypotheses.

Stressor as used in risk assessment can refer to anything that has the potential to affect the environment. The use of the term stressor can produce confusion because it is often used without any explicit definition. In risk assessment, it has most often been used as a synonym of the phrase "contaminant of concern." This usage implies a deleterious effect. In the field of ecology, a stressor does not necessarily cause a deleterious effect; in fact, it can often be a requirement for the maintenance of an ecosystem (e.g., fire). The use of the term stressor stimulated discussion and debate among the study team. The following brief discussion presents this background.

Stress has generally been perceived as negative and harmful. The pioneer work on stress by Hans Seyle (1973) shows that stress can be both a basic need and a harmful or even lethal influence. Seyle observed that stress is a constant influence (good or bad) in the lives of all organisms. He shows that stress-causing agents/activities, which he called stressors, initiate a "response" process in its receptors. The responses have three distinct phases: alarm, resistance, and exhaustion (Seyle, 1973).

The terms stress and stressor were used rather loosely in many scientific works in the past, generating confusion. This situation has been slowly shifting, as the mechanisms of stress are better understood. Currently they are defined as follows: stressors are any external pressures (chemical, physical, biotic, or environmental factors) exerted upon an ecosystem and "responses" or "stress" refers to the resulting biological effects (negative or positive). Subsidy stressors refer to those inducing a positive response. A good example is phosphorus, which can increase productivity and respiration in aquatic organisms such as algae. Contaminant/toxic stressors refer to those causing a deleterious effect in a community or ecosystem. Certain metals that have no nutritional value (e.g., mercury) constitute an example of a toxic stressor.

Seyle's original studies on stress and his "general adaptation syndrome" greatly influenced today's perceptions of stressors and response mechanisms. His theory stated that responses (readjustment or adaptation) depend largely on the intensity of the stressor(s), rather than on its nature (Seyle, 1973).

The *receptor* is the ecological entity exposed to the stressor such as organisms, populations, communities, and ecosystems (U.S. EPA, 1998). Ecological receptors in risk assessments are chosen using criteria such as sensitivity to the stressor, ecological importance, societal value; available knowledge about the biology of the receptor, and utility as a predictor of overall ecosystem health.

Conceptual models include risk hypotheses that describe predicted relationships among stressor, exposure, and response; and diagrams that illustrate the relationships presented in the hypotheses. Conceptual models describe how stressors interact with receptors and predict the responses to those interactions.

Risk hypotheses propose answers to questions about how exposure occurs, and predict the responses to the exposure. The predictions should be both testable and useful for generating research questions.

Conceptual model diagrams are visual representations of the risk hypotheses. Most often, they are flow diagrams that use boxes and arrows to illustrate relationships. They provide a dynamic representation of relationships and set the framework for final planning for the analysis phase.

The term *endpoint* has been used to describe two related but distinct concepts in risk assessment (i.e., assessment endpoint and measurement endpoint).

Assessment endpoints are the environmental values—the ecological entities and the desired attributes of those entities—to be protected. To be useful, they need to be ecologically relevant, susceptible to the stressors of concern, and relevant to risk management goals (U.S. EPA, 1998). Ecologically relevant endpoints reflect important characteristics of the system, and may occur at any level of organization—individual, population, community, ecosystem, or landscape. They often help sustain the natural structure, function, and biodiversity of an ecosystem. If, for example, planners want to protect an ecosystem where salmon spawn, the best assessment endpoint would be "reproducing populations of indigenous salmon," rather than some more vague endpoint, such as "viable fish populations."

There are three categories of *measures*. Measures of effect are quantitative expressions of an observed or measured effect of the stressor (formerly called measurement endpoints). Measures of exposure are measures of the pathways and amounts of exposure to stressors. Measures of ecosystem and receptor characteristics include ecosystem characteristics that influence the behavior and location of assessment endpoints, the distribution of stressors, and the life history characteristics of receptors. These measures increase in importance as the complexity of the assessment increases (U.S. EPA, 1998).

3.2 Review of Suggested Frameworks for the Study of Multiple-stressor Effects

Ecosystem stress theories are a useful framework for the evaluation of the effects of multiple stressors. A SETAC Pellston workshop on laboratory-to-field extrapolations held in May 1999 identified the usefulness of ecosystem stress theories as a framework for evaluating the capability of laboratory tests to predict effects in the field (Foran and Ferenc, 1999). One of the better-known models of ecosystem stress is by Odum (1985) (Table 3-1). Although Odum's predicted responses are based on the oversimplified assumption that undisturbed ecosystems develop along an expected successional path, they serve as a valuable tool for discussing and testing hypotheses regarding structural and functional (process-oriented) responses in disturbed ecosystems. More recent work has confirmed the validity of many of Odum's predictions for a range of ecosystem types (Rapport and Whitford, 1999). Therefore, the ecosystem stress model of Odum (1985) provides the foundation for our conceptual model.

The choice of structural or functional attributes of ecosystems as measures of stress is the subject of debate. Odum (1985) and others have argued that structural elements of ecosystems (e.g., community structure, biomass) are more sensitive indicators of anthropogenic stress than are process-oriented functional attributes, because species replacement and other adjustments keep the overall function of the ecosystem steady until the stress is overwhelming (Schindler, 1987). However, Pratt and Cairns (1996) challenge this view and cite examples demonstrating the sensitivity of functional ecosystem parameters.

Table 3-1. Trends Expected in Stressed Ecosystems, from Odum (1985).

Ecosystem Property	Expected Trend
Energetics	
Community respiration	Increase
P/R (production/respiration)	Becomes unbalanced (< or > 1)
P/B (production/biomass) and R/B (respiration/biomass)	Increase
Importance of auxiliary energy	Increase
Exported or unused primary production	Increase
Nutrient Cycling	
Nutrient turnover	Increase
Horizontal transport	Increase
Vertical cycling	Decrease
Nutrient loss	Increase
Community Structure	
Proportion of r-strategists	Increase
Size of organisms	Decrease
Lifespan of organisms	Decrease
Food Chains	Shorten
Species diversity	**Decrease**
Dominance	**Increase**
General	
Inputs/outputs to and from other ecosystems	Increase
Autogenic successional trends	Reverse
Efficiency of resource use	Decrease
Parasitism and other negative interactions	Increase
Mutualism and other positive interactions	Decrease
Functional properties are more robust than structural	
Properties such as species composition	

The U.S. EPA's SI guidance document (U.S. EPA, 2000) identifies stressors causing biological impairment in aquatic ecosystems, and provides a structure for organizing the scientific evidence supporting the conclusions. The ability to accurately identify stressors and defend the evidence supporting those findings is a critical step in developing strategies that will improve the quality of aquatic resources.

The SI process put forward by U.S. EPA is prompted by biological assessment data indicating that a biological impairment has occurred (U.S. EPA, 2000). The general SI process entails critically reviewing available information, forming possible stressor scenarios that might explain the impairment, analyzing those scenarios, and producing conclusions about which stressor or stressors are causing the impairment. The SI process is iterative, usually beginning with a retrospective analysis of available data. The accuracy of the identification depends on the quality of data and other information used in the SI process. In some cases, additional data collection may be necessary to accurately identify the stressor(s). The conclusions can be translated into management actions and the effectiveness of those management actions can be monitored.

A SETAC Pellston workshop held in 2001 (Baird and Burton, 2001) dealt with multiple stressors in an ecological context. The workshop examined tools and approaches that: identify and separate multiple stressors within ecosystems; link individual responses to stress within ecosystems; examine the role of environmental variability in evaluating stressor effects; and assess multiple stress impacts in food webs.

Several authors have proposed that any evaluation of the effects of multiple stressors on stream ecosystems should use a weight-of-evidence approach. This approach is consistent with guidance provided by Suter (1993) and is also consistent with the U.S. EPA (1988) framework for ecological risk assessment.

Lowell et al. (2000) have found three steps to be particularly critical in the development of a weight-of-evidence approach for ecological risk assessment: 1) establishing causality, 2) defining acceptable change, and 3) linking environmental components in a decision-making framework. The authors assert that distinguishing among the cumulative affects of multiple stressors (which sometimes have confounding effects) requires establishing definitive causality links to each stressor. Establishing a strong causal link requires a weight-of-evidence approach that combines information from a variety of sources. The strength of the causal link can be evaluated by using a formal set of criteria originally developed in the field of epidemiology. These criteria are outlined in the first column of Table 4-3.

Defining significant change is a required element of any risk assessment, and is a fundamental component of any multiple stressors study design. Magnitudes of the observed effects must be compared against a predetermined critical effect size. If the measured effect exceeds the critical effect size, it is judged to be ecologically significant. There are a number of suggested methods for determination of critical effect sizes (e.g., Washington State Department of Ecology, 1991; Elliott, 1996; Lowell, 1997; Hope, 1998). The methods can include information from previously collected data, such as the normal range of variability seen in reference communities. Although these studies focus on the ecological component of risk assessment, critical effect sizes can also be influenced by other factors, such as societal values.

The study design developed by this project is based on the framework developed by Odum (1985), but also incorporates risk assessment methods and terminology, including the use of a weight-of-evidence approach. The conceptual model and risk hypotheses arise from ideas generated by the Odum framework. The suggested approach to testing the risk hypotheses follows the discipline imposed by progressing through the problem formulation stage of risk assessment. The proposed experimental method makes use of several lines of evidence using experimental and observational tools in an integrated manner. The suggested approaches for assembling and interpreting the weight of evidence are derived from the U.S. EPA SI process, as well as from the experience of the study team (e.g., Lowell et al., 2000).

3.3 Bridging the Gaps in Our Knowledge of Multiple Stressors

The 1997 Pellston Multiple Stressors workshop (Foran and Ferenc, 1999) attempted to assess the state of knowledge on incorporating multiple stressors in ecological risk and impact assessment and to develop a framework as well as specific methods and models for the qualitative and quantitative evaluation of multiple stressors. The conceptual model developed by the study team provides a framework for predicting the interactions between stressors and their effects on habitat, structure, and function within an ecosystem, based on different baseline conditions.

Baseline conditions refer to the existing ecological conditions in a particular habitat before the addition of any specific stressor(s). For example, the baseline conditions in a river may refer to the existing concentration of nutrients and certain contaminants, temperature, DO, and other parameters. Baseline conditions may also include population densities and species composition of benthic invertebrates, fish, and plankton present in that river.

The conceptual model design was developed using the following questions to stimulate discussion:

- Can stressors be both subsidies and toxicants (following the ideas of Seyle)?
- Can interactions between stressors be predicted?
- Do structural or functional responses occur first?
- How do habitat changes interact with other stressors?
- Would responses to stressors be similar under different baseline conditions?

A set of working hypotheses was developed that predict how multiple stressors might affect ecosystems under various baseline conditions and are the central component of the conceptual model.

3.4 Fundamental Characteristics of the Conceptual Model

The conceptual model originally was proposed to address types of ecosystem response to stressors, relationships among stressors and responses; cumulative effects of stressors, and recovery processes following removal of one or more stressors.

The experimental approach would have to incorporate an extended test period (including winter months), which posed significant logistical problems. So, it was determined by the PSC and the study team that the inclusion of recovery processes would not be practical at this stage. The recovery hypothesis may be included after the first round of testing using the conceptual model and proposed study design approach.

The conceptual model for multiple-stressor effects on ecosystems must be a flexible and scientifically defensible tool that has a strong experimental emphasis. It was also agreed that the conceptual model would build on existing observational databases; generate hypotheses that can be tested experimentally; use formal hypothesis testing to evaluate its usefulness; and relate experimentally obtained results to field observations, allowing for model validation.

3.4.1 Concept of ì Habitatî

Anthropogenic changes to habitat can be very important, if not dominant, stressors in ecological systems. Habitat is broadly defined as the inter-connection between the chemical, physical and biological properties of the location where a particular biological community lives (Figure 3-1). Although this broad definition is useful for understanding the interactions that occur to produce conditions required by a particular species, it is difficult to incorporate into a conceptual model for multiple stressors because it includes the potential for several different stressors within the "habitat" category. Therefore, for the purposes of the conceptual model and study design in this report, habitat is more narrowly defined as the physical characteristics of the environment (e.g., substrate type or water flow) required for the successful reproduction, development and growth of a species.

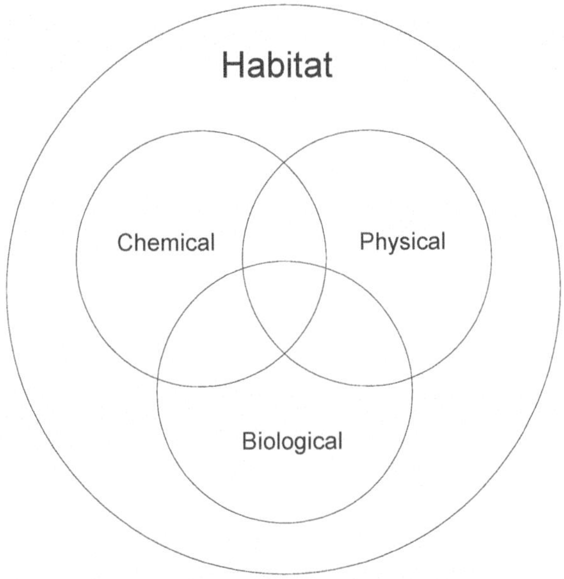

Figure 3-1. Schematic Diagram of the Broad Definition of Habitat.

The physical environment, or habitat, of stream ecosystems exerts substantial control over population abundance and hence, community composition. Physical factors of particular importance to lotic organisms are DO concentrations, water temperature, substrate composition, discharge and hydrodynamics, and channel geomorphology.

All aerobic organisms must obtain sufficient oxygen to drive their metabolic machinery. In most unaffected streams, DO concentrations generally are at or near saturation because of turbulent mixing and thus, oxygen rarely limits biological diversity. Under certain conditions, however, oxygen depletion may occur in particular habitats or at certain seasons or times of the day. Inputs of chemical or physical waste into streams result in chemical oxygen demand or BOD that can deplete oxygen downstream of inputs. High primary production resulting from, for example, nutrient input and riparian alteration can infuse water with oxygen during the daytime but deplete oxygen during the night when respiration dominates. If coupled with increased water temperature (see below), low DO can be particularly stressful to aquatic organisms.

Water temperature imposes considerable constraints on all aquatic organisms. For example, temperature affects animals directly by regulating their metabolic rates and thus development from egg to adult, and indirectly, by influencing such things as fluid dynamics, gas saturation constants, and primary production rates. Water temperature fluctuates both spatially and temporally in the river systems. Some animals prefer a narrow range of temperature; others can tolerate a broader range of temperature. Consequently, temperature fluctuation can limit the types of animals found within a habitat and any significant change in the water temperature will likely alter species composition. Water temperature is often coupled with riparian condition, because reduced canopy cover results in increase in irradiance leading to altered water temperature at both extremes.

At the scale of local habitat, substrate composition, and hydrodynamics are probably the most important factors determining the types and abundance of benthic organisms present, especially aquatic macroinvertebrates (Lamberti, 2002). The stream substrate has obvious importance because the vast majority of stream macroinvertebrates spend most of their lives attached to substrates and many stream fishes (and most fish eggs) are associated with the substrate. Lotic substrates can be broadly divided into inorganic substrates (geologic material ranging from silt to boulders) and organic substrates (fine organic particles up to logs). The particle size of inorganic matter has a large influence on community structure. Coarser bed material (e.g., gravel, cobble, and boulders) generally provide more and better interstitial habitat for macroinvertebrates than do finer sediments (e.g., sand and silt). Fine sediment inputs to streams are problematic because the tight packing of particles restricts physical habitat and the trapping of detritus, and can limit the availability of oxygen. Therefore, a lower diversity of macroinvertebrates is typically found in fine sediments, and developing fish eggs can be deprived of oxygen in habitats with significant proportions of fine sediments.

Variation in streamflow (from floods to desiccation) is a major factor responsible for biological structure (Lamberti, 2002). Streamflow variation is a natural feature of all streams and rivers. But human disturbance of rivers, riparian zones, and watersheds has altered the natural flow regime of most rivers. Lower baseflows (caused by water withdrawal), higher peak flows (caused by impervious surfaces), or reduction in high flows (caused by impoundments) are but a few of the effects that humans impart on river systems. These changes can interact with the other physical features described above. Hydrodynamics (i.e., smaller-scale water movements) affect many aspects of animal biology—including body form, food acquisition, and movement—and have a large influence on periphyton abundance and composition. Near-bed hydrodynamics, such as shear, are closely tied to discharge, and channel form. Channel simplification often results in higher shear and therefore greater hydraulic stress on benthic organisms.

Channel geomorphology is the physical form that river systems assume and is a dynamic interaction between the tendency of rivers to carve new channels (endogenous control) and the river valley to constrain such improvements (exogenous control) (Lamberti, 2002). Humans have greatly increased the degree of exogenous control on river meandering by confining the areas through which rivers are allowed to flow. This control is apparent in all landscapes that have been modified by humans, but especially in urban and agricultural settings. Humans have also greatly simplified in-stream morphology by simplifying channels to expedite water transport. In general, meandering streams have been channelized; roughness elements such as large wood have been removed to accelerate drainage. Where riparian forests have been harvested, the future source of wood debris has been terminated. The alterations combine to diminish the diversity of habitat available for stream biota and to greatly restrict the potential productivity of river systems.

3.4.2 Types of Stressors

Stressors can be physical, chemical, or biological. Physical stressors are those factors that affect the physical attributes of habitat required by aquatic species, and include such variables as discharge, water temperature, irradiance, sediment composition, and channel geomorphology. Normally, it is the extremes in these variables that negatively affect organisms, such an extreme high or low flow, high temperature, excess ultraviolet radiation, fine sediment inputs, or simplified channel morphology. Chemical stressors include inorganic and organic chemical substances introduced into the watercourse. Biological stressors include pathogenic organisms and exotic species.

The conceptual model presented in this report focuses on physical and chemical stressors. Inclusion of biological stressors into the overall conceptual model can occur in the future after the overall usefulness of the conceptual model for prediction of effects of physical and chemical stressors has been tested.

As explained in Section 3.3, a stressor can be beneficial (a subsidy) or harmful (a toxicant). Physical variables also can function as subsidies or toxicants. For example, increased irradiance caused by riparian disturbance can be a subsidy and stimulate food web production from the bottom-up; however, excess light can be a toxicant by producing harmful algal blooms whose respiration can deplete DO at night. Similar dual roles, depending on degree of modification, can be played by other physical factors including water temperature, discharge, and fine sediments. For example, a modest increase in water temperature that can result in higher growth rates of invertebrates or vertebrates without inflicting mortality. In general, simplification of stream geomorphology has negative effects on biological diversity and productivity, although certain taxa may benefit from habitat modification.

Some stressors produce hormetic effects at low concentrations. Hormesis is defined as low-dose stimulation followed by higher-dose inhibition (Calabrese and Baldwin, 1998). Many minerals and vitamins are toxic for organisms at high exposures but small amounts can give better performance or fitness than at zero exposure (Parsons, 2002). Many hormetic agents are a normal component of habitats including various metals that are exceedingly toxic at high concentrations. Background hormesis may be almost universal for substances present through geological time, such as mercury (Parsons, 2002). Evidence of chemical hormesis was judged to have occurred in approximately 350 of the 4,000 studies evaluated by Calabrese and Baldwin (1998). Chemical hormesis was observed in a wide range of taxonomic groups and involved agents representing highly diverse chemical classes. Hormetic responses were generally observed to be of limited magnitude. The average low-dose maximum stimulation was approximately 50% greater than controls. The hormetic dose-response range was generally limited to about one order of magnitude, with the upper end of the hormetic curve approaching the estimated no observable effect concentration for the particular endpoint.

Table 3-2. Examples of Stressors by Type.

Stressor Type	Subsidy-Toxicant Stressors Depending on Concentration or Dose	Toxicant Stressors with No Known Subsidy or Hormetic Effect
Chemical		
Nutrients	Nitrogen	
	Phosphorus	
	Organic Carbon	
Metals	Zinc	Lead
	Chromium	Thallium
		Uranium
Organics	Alkyl sulphate and other surfactants	Dioxins and furans
	Herbicides	Polycyclic aromatic hydrocarbons
Physical		
	Heat	
	Flow rate	
	Radiation (light)	
	Suspended solids	

3.5 The Schematic Conceptual Model

The proposed conceptual model illustrated by the linkage diagram in Figure 3-2 encompasses the following key assumptions:

- Interactions between the stressor and the ecosystem depend on the existing baseline conditions;
- For each stressor/baseline combination there will theoretically be a functional and a structural response;
- Effects on habitat are caused by structural or functional responses to stressors; and
- Effects on habitat can cause structural or functional responses.

Figure 3-2 shows the chain of events (solid arrows) that a particular ecosystem will undergo under the influence of a stressor(s): functional and/ or structural responses (that may interact with each other) are triggered, which in turn affect the original habitat and vice versa. The figure illustrates the influence of baseline conditions on the ecosystem responses (dotted lines).

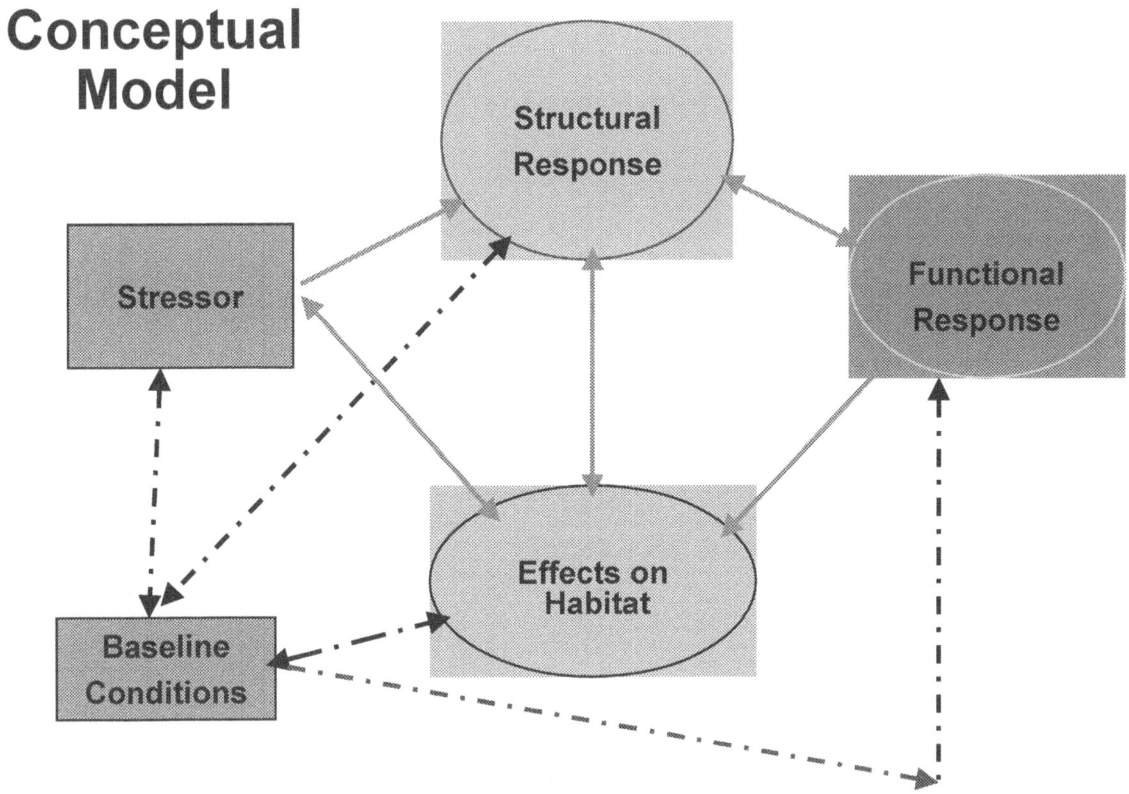

Figure 3-2. Schematic Representation of the Interactions as Predicted by the Conceptual Model.

The following two examples illustrate how the conceptual model encompasses stressors acting both as subsidy and toxicant.

In Example 1A (Figure 3-3), the stressor is a nutrient, such as phosphorus (P) and the baseline or background condition is a nutrient-poor habitat. The expected responses would be:

- Functional response: increase in productivity and respiration rates;
- Structural response: initial increase in abundance, up to a stressor threshold concentration (subsidy stage); and
- Effects on habitat: positive effects with increased habitat for various species including for fish.

In Example 1B (Figure 3-4), the stressor is again the nutrient phosphorus (P), which is present at very high concentrations, and the baseline or background condition is a nutrient-poor habitat. Expected responses include:

- Functional response: production/respiration is unbalanced; respiration rates increase, export of unused primary production.
- Structural response: decrease in species richness, increased abundance, and prevalence of smaller organisms (e.g., blue-green algae replace green algae).
- Effects on habitat: a negative effect during the toxic stage because of decreased DO and from algal decomposition and decreased light caused by shading from algal blooms.

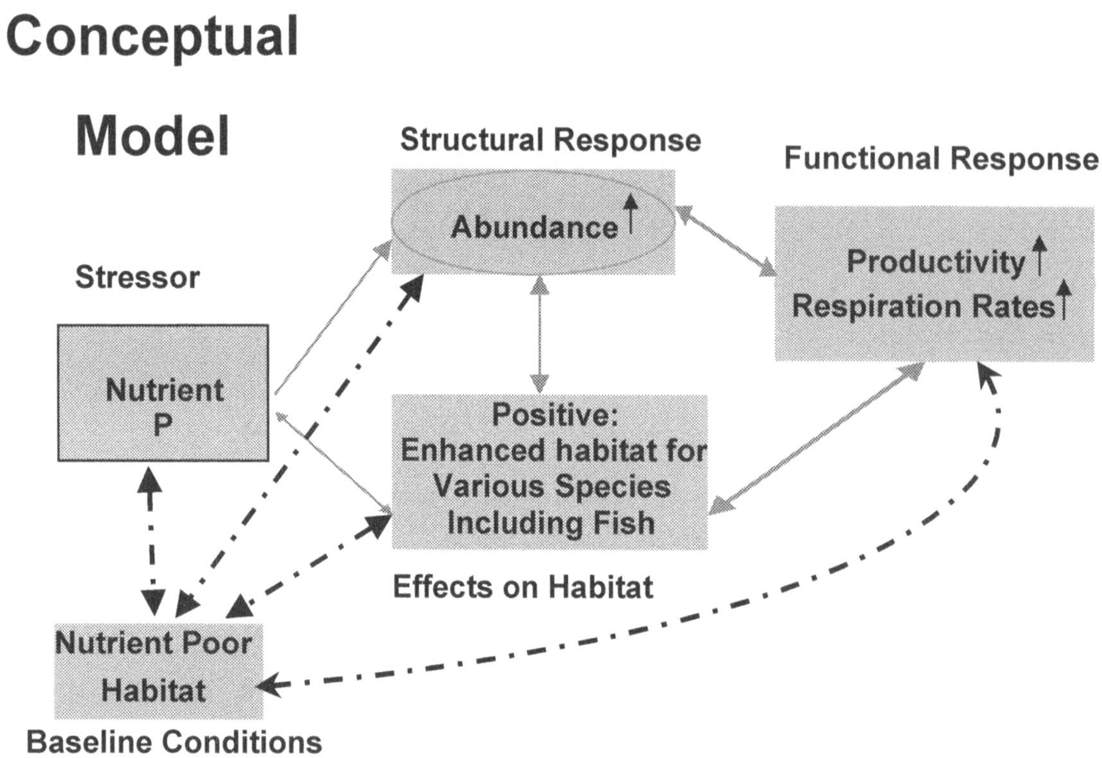

Figure 3-3 . Example 1A: Input (P) is a Subsidy.

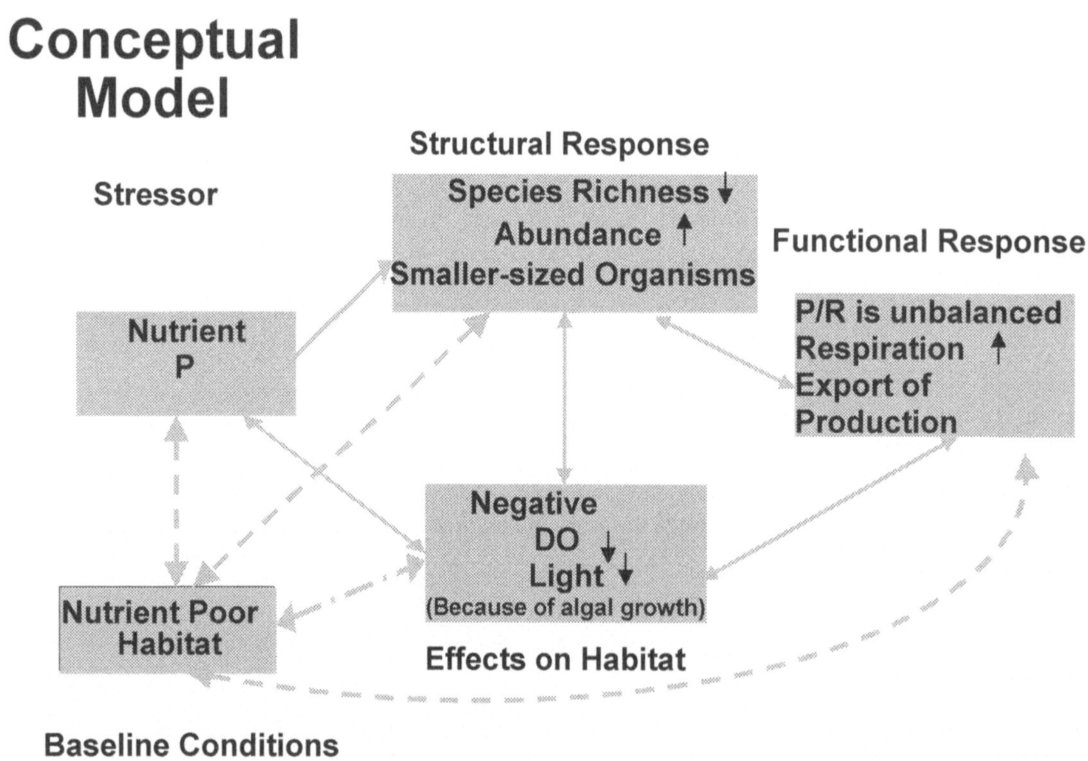

Figure 3-4. Example 1B: Input (P) as a Toxicant.

3.6 Risk Hypotheses

3.6.1 Primary Background for the Risk Hypotheses

Structural Versus Functional Response

In ecosystems, there are two general categories of response to stressors: structural and functional. Either type of response can occur through changes in natural habitat and each type of response can alter habitat.

Odum (1985) hypothesized that effects on ecosystem structure will occur before detectable changes in function caused by the functional redundancy present in most aquatic ecosystems. Toxicant stressors will remove the most sensitive species first. The sensitive species will rapidly be replaced by other, more tolerant species, maintaining ecosystem function. Thus, functional responses may initially be dampened at lower stressor concentrations. This effect starts at the lowest trophic levels of the community (e.g., blue-green algae replacing green algae). Odum (1985) suggests that structural changes would include an increase in r-strategists (small, fast-growing, short-lived, ruderal species) and a decrease in k-selected organisms (large, long-lived, slow-growing species) accompanied by shorter food chains and a decrease in species diversity. The initial structural responses to a toxicant can create a cascade of other responses, such as disruption of biotic interactions that structure communities (e.g., competition, predation, mutualism and parasitism, and a resulting increase in temporal variance in population densities because of decreased biotic control).

Odum's concept of structural change always preceding functional change was challenged by the study team once it was agreed that nearly all stressors can be either a subsidy or a toxicant, depending on concentration and on the ecosystem's baseline conditions. The study team then progressed to the idea that when a stressor is a subsidy stressor, functional responses can take place first in the system (e.g., increase in productivity).

Functional responses would occur before structural change when a stressor is present in concentrations where it does not impair the system. The lack of impairment may indicate a true subsidy or hormesis. Although a true subsidy refers to a measurable, lasting enhancement of the system's overall function, a hormetic response may result from the temporary overcompensation (or repair mechanism) to some inhibitory challenge.

The coral reef studies conducted by Odum (1985) provide a good example of the hormetic effect that can mimic a true subsidy. Odum observed an increase in production-to-biomass (P/B) or maintenance-to-biomass (R/B) ratios and an imbalance in the production-to-maintenance ratio (P/R) ratio from >1 to <1 when any input of toxic substances (e.g., cadmium) or silt was added. These changes were attributed to diversion of energy to favor production or respiration.

As concentrations of a stressor increase, the stressor will reach a point where it shifts from a subsidy to a toxicant stressor. At this point, it is expected that structural changes will occur first, followed by functional changes such as:

- ♦ An increase in respiration/maintenance of communities.
- ♦ P/R ratio becomes unbalanced (>1 < 1).
- ♦ P/B or R/B increases.
- ♦ Auxiliary energy sources increase.
- ♦ Export or unused primary production increases.

According to Odum's 1985 model, if stressor(s) concentration/intensity is severe enough, an ecosystem can eventually be pushed into a state of "alternate stability," where it loses its ability to recover, even when the stressors are removed. Known examples are extreme physical stressors (e.g., siltation or channelization of streams) and biological stressors (e.g., introduction of exotic species, such as zebra mussels).

Subsidy Versus Toxicant Stressors

The ecosystem subsidy-stress model of Odum et al. (1979) is one of the best-known models for predicting the response of ecosystems to stressors and provided additional theoretical background for the conceptual model presented in this report. A key to Odum's approach is the assumption that a single major stressor has disrupted the ecosystem from its nominal state, thereby producing either a subsidy or stress response. If the stressor is stimulatory of growth or production, like heat or nutrients, then a subsidy effect may be observed at low levels. Overall, the subsidy side of the equation in Odum's model has been given less consideration than the stress side. Nevertheless, scattered observations in the literature broadly support Odum's conception. Additions of nutrients, organic matter, and energy (heat or light) are stimulatory at the species, community, and ecosystem levels when the amounts added are small. These additions tend to become stressors by indirect routes when the levels rise above certain minimum thresholds. For example, increasing levels of nutrients will continue to increase plant production (and hence animal production) until they are no longer limiting to growth. The disruptive effects of high-nutrient subsidies arise from their unbalancing of competitive relationships, wherein a few aggressive, fast-growing species are favored over the large number of specialized, slow-growing species, and from such indirect effects as smothering, loss of habitat and de-oxygenation during decomposition.

Toxic chemicals may act as subsidies to the ecosystem if the concentration is low enough. Toxic organics contain reduced carbon and therefore are a food source for bacteria; increased microbial production is common in microcosms or natural systems dosed with chronic low doses of toxic organic compounds such as pesticides. Even heavy metals can be stimulatory at low doses. Members of the chironomid subfamily Orthocladinae often thrive under chronic metal exposures that are toxic to other insects (La Point et al., 1984; Leland et al., 1989; Kiffney and Clements, 1994).

The subsidy-stress model is based on observational studies and assisted the study team in the early iterations of the conceptual model because it helped confirm the fact that a stressor could be either a subsidy or a toxicant. While very useful, the subsidy-stress model has some limitations:

- ♦ The model does not acknowledge community-level changes such as species replacement and diversity alterations, which would occur well before the cusp between subsidy and toxicant. Community-level changes account for the gradual change from subsidy to toxicant.

- Inaccuracy in the shape of the curve representing the immediate effect of a toxicant on an ecosystem-level response because it does not account for the dampening of responses in production, decomposition or nutrient cycling.
- The curves representing ecosystem and community responses are sigmoidal and imply that severe stress will have an enormous affect on the species present. That is unrealistic in the case of severely compromised ecosystems, which are already extremely simplified and where only stress-tolerant hardy species prevail.

Interactions Among Stressors

Understanding the nature of interactions between toxicants and subsidies is key to the prediction of integrated ecosystem response to multiple stressors. Nutrient or energy subsidies lead to an acceleration of metabolic processes. This acceleration, in turn, can exacerbate or reduce the response to other stressors: fast-growing organisms absorb toxic heavy metals more quickly (e.g., Wang and Dei, 2001), but may be more resistant to toxicity because of greater vigor and dilution of toxicants in a larger mass of tissue. All other things being equal, the same effect should apply to subsidies of all kinds. To the extent that this effect is generally true, it may be possible to predict the combined effect of stressors and subsidies using a simple additive model in which subsidies are given negative weight.

3.6.2 The Risk Hypotheses

The study team reviewed and developed the conceptual model through internal discussions and meetings with the PSC. The new approach emphasizes that almost every substance (stressor) is potentially a subsidy or a toxicant to an ecosystem, depending on its concentration. The conceptual model put forward is founded upon three basic risk hypotheses:

- Subsidy stressors first cause a functional response.
- Toxicant stressors first cause a structural response.
- Multiple stressors will act in an additive manner.

On the basis of experience with well-known subsidies such as heat and nutrients, the first hypothesis assumes that functional response to subsidies can be almost instantaneous. For example, heat produces a rapid increase in metabolism. The addition of a limiting nutrient produces a rapid response in primary production of algae.

The second hypothesis assumes that toxicants will change structure first because of the rapid reduction or elimination of sensitive species. This would be particularly true if acutely lethal thresholds were exceeded.

The third hypothesis assumes that when faced with two or more stressors, population declines will happen more swiftly, species replacements will be accelerated and loss of species will begin at an earlier threshold of each stressor than for either stressor alone. Synergistic or antagonistic responses are also possible but are likely to be less common than simple additivity.

The effect of multiple stressors on ecosystems will depend on the net response at the population and community levels. Consequently, in the face of two or more stressors, dampening and compensation mechanisms would be overcome sooner, and the response curve at the ecosystem-level would be steeper than for either stressor individually.

An additional important consideration for each of the above three hypotheses is that the responses to different stressors will invariably depend upon the existing baseline (or background level) conditions.

The subsidy-toxicant curve (Figure 3-5) illustrates the concept of a continuum in response to a stressor. The subsidy-toxicant curve is divided into different sections labeled [a], [b], [c] and [d]. Each section corresponds with a different concentration/response combination.

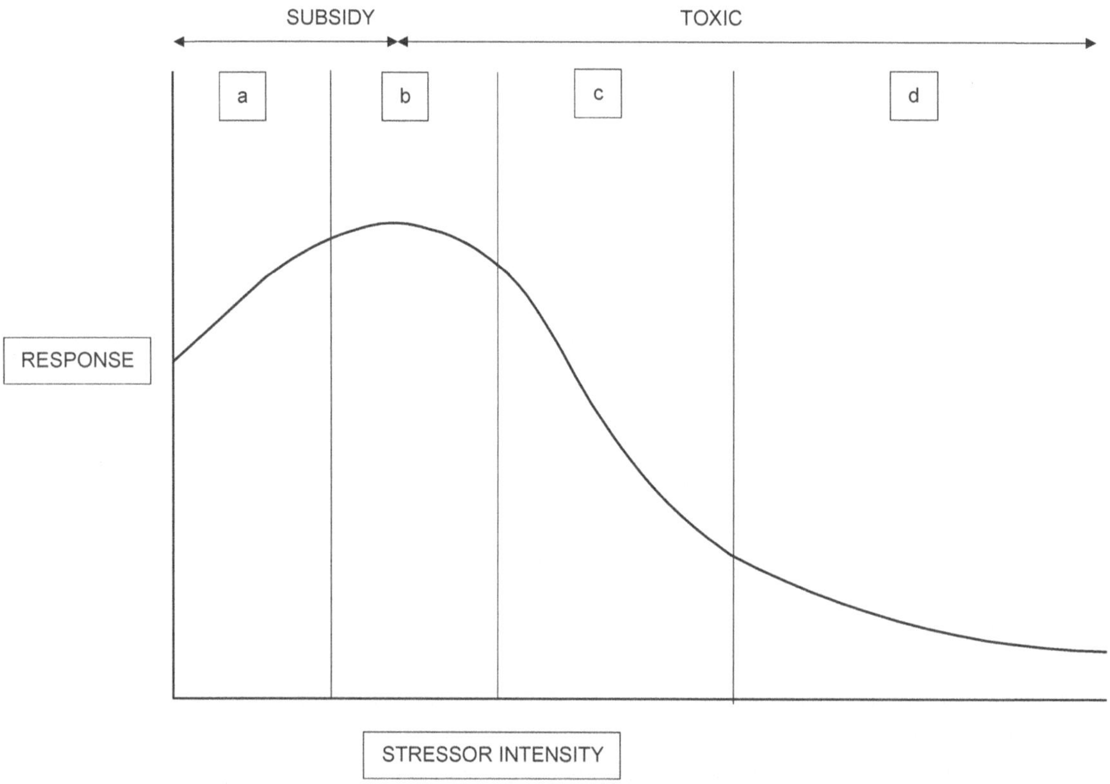

Figure 3-5. Subsidy-Toxicant Curve.

For a limiting nutrient, or an addition of energy such as sunlight or reduced carbon, the *y* axis might be positioned at [a]. Small or moderate increases in the stressor cause a commensurate positive ecosystem response (subsidy). Only after a considerable increase in dose or intensity does the effect become negative (toxicant), until eventually the curve flattens out when only a few hardy species survive. The subsidy portion of the proposed curve can occur as a result of:

- Clear stimulatory effect from a substance(stressor) acting as a subsidy; or
- A form of "hormesis," or hormetic response.

Many toxicants would be modeled by positioning the curve at [b]. Over most of the concentration range, the effect of a toxicant is negative, producing a proportional negative ecosystem response with increasing concentration. Very low doses, however, may be stimulatory or produce an over-compensation leading to a positive ecosystem response over a narrow dose range.

Substances that are of no biological utility (such as lead) may produce a stressor-response curve beginning at [c]. For these substances, there is no stimulatory effect no matter how small the concentration, because organisms cannot use the material as an energy source or structural component. Lead and cadmium are often cited as examples, but the effects of very small lead additions have not been examined extensively. There is limited evidence that low concentrations of cadmium actually induce a stimulus to algae. Among the organic compounds, it is possible that some of the most recalcitrant substances such as polychlorinated biphenyls (PCBs) or dioxins are toxic at all concentrations, but for virtually all other organic substances, low concentrations are stimulatory because they provide an energy source for bacteria. This is certainly true of ordinary pesticides such as 2,4-D and all the substances we normally consider components of dissolved organic matter. In cases where hormetic responses occur, such as those observed by Odum's coral reef studies (Odum, 1985), then the stressor-response curve would begin at [b].

Finally, adding a stressor to an already much-stressed ecosystem would be equivalent to positioning the *y* axis at [d]. The stressor would still cause a negative ecosystem response, but less so than an equivalent dose or intensity in an unstressed ecosystem because the stressed system has already lost its sensitive components.

The proposed conceptual model is testable and can be used to predict ecosystem responses to multiple stressors of different kinds, which is the ultimate objective. The model combines the subsidy/toxicant dichotomy into a single relationship, addresses a long-standing controversy about whether functional or structural changes are expected first for any given situation, and leads to interesting new hypotheses. For example; if a substance is of type [c], does this imply that ecosystem response is already less than a theoretical maximum that would occur if all traces of the substance were absent from the environment? And does that suggest that "natural" toxicity may be at work in some places?

The unifying idea in this model is the hypothesis that the subsidy/toxicant curves for different kinds of stressors are all of the same general shape; the difference is where on the curve the new input places the ecosystem in question. This differs from Odum's (1979) model, in which he envisioned subsidy and toxicant curves as different. Hence, nutrients like phosphorus added to an oligotrophic ecosystem begin far to the left of the curve and produce a short-term productivity increase before there is any loss of species. For a strong toxicant like copper, the curve is shifted to the right, so the toxicity response begins almost at once.

Practical Application of the Risk Hypotheses

Hypotheses 1 and 2 have significant implications regarding how receiving environment monitoring data are collected and interpreted. Determination of threshold concentrations for when a stressor shifts from a subsidy to a toxicant will provide a very important basis for interpretation of the need for risk management (refer to Figure 3-5). If monitoring data show that a stressor or combination of stressors is well within the subsidy portion of the curve, risk management measures are not indicated. On the other hand, if monitoring data show that a stressor or combination of stressors is approaching or exceeding the threshold where significant negative ecosystem responses occur, then planning for risk management measures to maintain concentrations in the subsidy portion of the curve would be warranted.

Risk Hypothesis 3 is the most relevant to decision-making regarding multiple stressors because it can provide answers to the key question: "Are multiple stressors acting in an additive manner?" Current regulatory practice assumes simple additivity of deleterious effects. There is no accounting for subsidy effects or for compensatory mechanisms, nor is there any accounting for potential synergistic effects (both positive and negative). In practice, it is extremely important to understand whether multiple stressors are, indeed, producing additive effects or whether some stressors dominate in terms of determining the overall ecosystem response. Results of experiments that test hypothesis 3 can be used to rank stressors so managers can respond specifically to the stressors that produce the dominant toxicant-type effects in the receiving environment. Decision-makers are currently meeting their pollution control targets, such as total maximum daily loads (TMDL), based on additive measurements. However, it is not known whether meeting these TMDL targets is truly protective of the receiving environment or whether it is significantly under- or over-protective, thereby producing a misallocation of risk management resources.

Testing the conceptual model and risk hypotheses will produce a growing body of knowledge useful for decision-making. Results will help identify maximum stressor concentrations that a particular ecosystem can withstand from all sources, point and nonpoint source contributions, before deleterious effects start to take place.

Final judgment on an effluent's maximum allowable concentration of a stressor(s), however, must be based on the total input of such a stressor to the ecosystem in question. For example, for stressors that are nutrients (particularly phosphorus) it is perhaps just as important to determine the total nonpoint source contribution to the system. In such cases, the use of a mass balance may be essential.

Examples of the Conceptual Model and Risk Hypotheses

The study team worked through several examples of combinations of stressors to test the utility and flexibility of the conceptual model. Selected examples are presented below.

Examples 1A–B, 2A–C, and 3A–C (Figures 3-5 through 3-13) illustrate predicted responses in ecosystems from two chemical stressors: copper and phosphorus. Three concentrations of each stressor are used; zero (representing no change from baseline), low, and high. The assumed baseline conditions are an "undisturbed environment" with background concentrations of copper and an oligotrophic nutrient status with very low background phosphorus concentrations. The "low" concentration of copper is assumed to be less than the lowest observed effect concentration (LOEC) for sensitive algal species. The "high" concentration of copper is assumed to be above the LOEC, but not above the lethal concentration (LC50) for most algal species. The "low" concentration of phosphorus is assumed to be in the range associated with mesotrophic (moderately enriched) conditions. The "high" concentration of phosphorus is assumed to be in the range associated with eutrophic (highly enriched) conditions. Predicted responses are for primary producers (algal species) only. The predicted changes in ecosystem function and community structure are based on Odum's (1985) expected trends in stressed ecosystems (Table 3-1), combined with the study team's understanding of subsidy/hormetic effects.

Figures 3-5 to 3-13 use examples of combined stressors to illustrate the responses predicted by the conceptual model. Part A of each figure uses the subsidy-toxicant curve to illustrate these effects (see shadowed portions of the curve) and Part B of each figure uses the linkage diagram of the conceptual model to illustrate predicted responses (structural and functional) and the effect on habitat.

Example 1A:
Stressors: [Cu] = Low; [P] = 0
Baseline Conditions: undisturbed, oligotrophic environment

Part A: Stressor-Response Curve

The only stressor in this example is the chemical stressor copper in "low" concentrations. The model predicts that toxicant stressors present in low doses where baseline copper concentrations are at background may produce a stimulatory or a hormetic response. Because the threshold for toxicity of copper to algae is so low, a small addition of copper to background concentrations may increase the concentrations to a point approaching or slightly exceeding the toxicity threshold. However, the model predicts that most of the ecosystem response would occur within the subsidy portion of the curve.

Part B: Linkage Diagram

The linkage diagram illustrates the predicted ecosystem responses to copper in low concentrations, where baseline copper concentrations are at background. The overall response is expected to be in the subsidy region of the subsidy-toxicant curve; therefore, a functional response is expected to dominate. The functional response is predicted to be a stimulation of growth in algae caused by hormesis. Many metals, including copper, have been shown to have hormetic effects at low doses. The magnitude of stimulation may be quite small. Calabrese and Baldwin (1998) report that in most cases the magnitude of the stimulatory response is 30% to 60% greater than controls. Structural responses are predicted to be nondetectable because copper concentrations will remain at or below the toxic threshold. There may be small shifts in community structure because of the variable sensitivity of algal species to copper; some species may show a pronounced stimulatory response whereas other highly sensitive species may be slightly inhibited. However, these small shifts would be well within the range of natural variability. Therefore, there would be no detectable change in habitat conditions caused by changes in algal community structure.

(A) SUBSIDY-TOXICANT CURVE

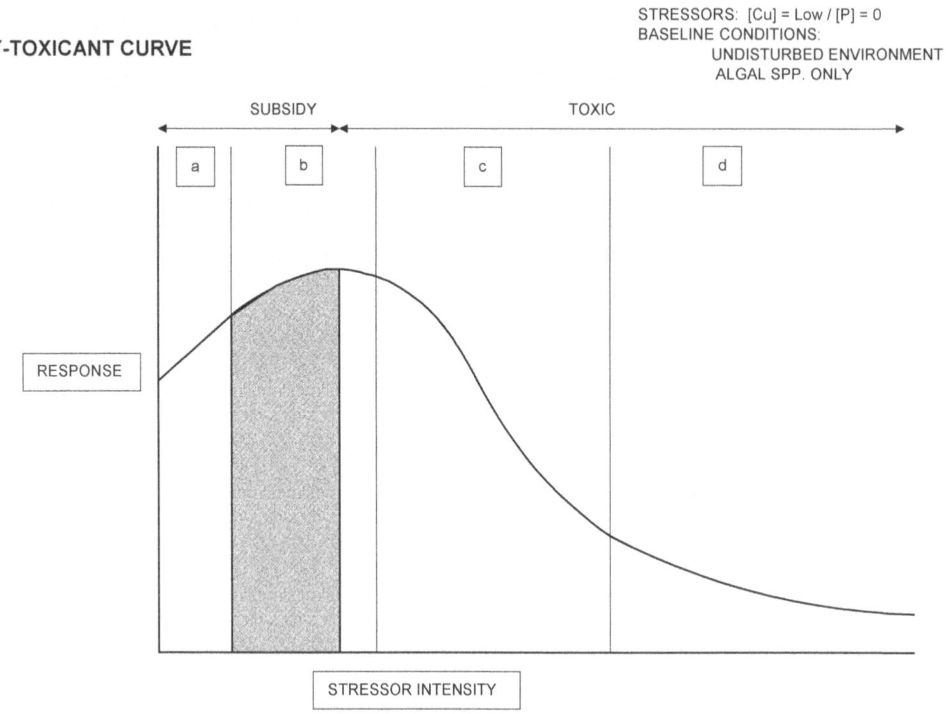

(B) LINKAGE DIAGRAM

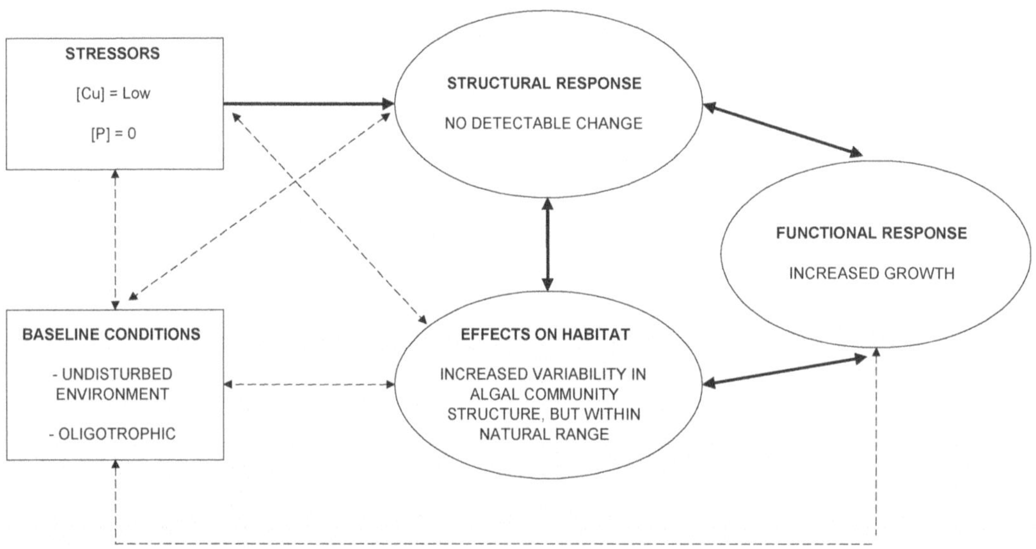

Figure 3-6. Conceptual Model Example 1A.

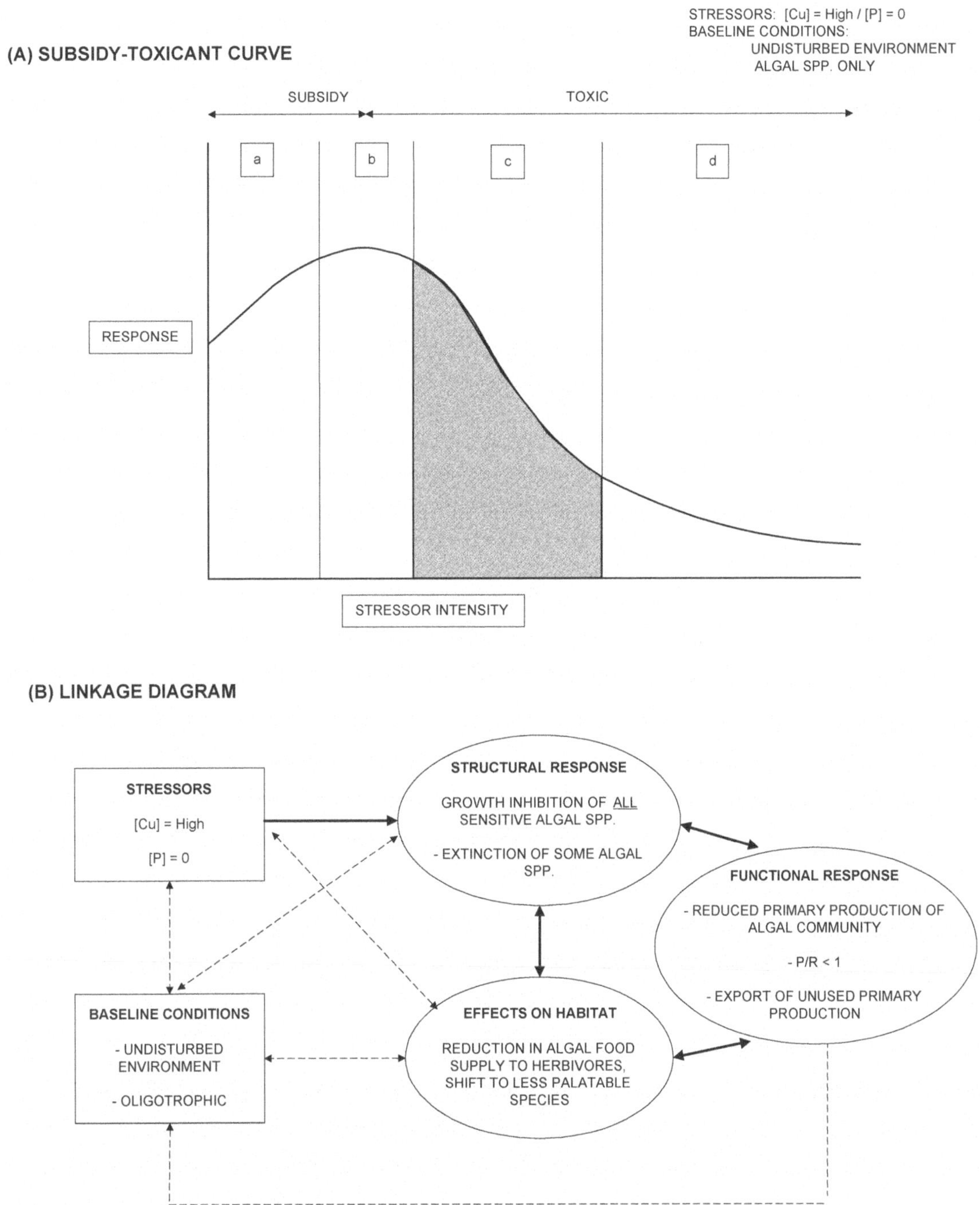

Figure 3-7. Conceptual Model Example 1B.

Example 1B:
Stressors: [Cu] = High; [P] = 0
Baseline Conditions: undisturbed, oligotrophic environment

Part A: Stressor-Response Curve

The predicted response when copper concentrations are high lies within the toxicant portion of the subsidy-toxicant curve [c]. Because the baseline copper concentrations are at background levels and the definition of "high" concentrations is greater than the LOEC but not greater than the LC_{50}, the total copper concentrations would not fall in the part of the curve where acutely lethal conditions would exist for many species (past the [d] point). Therefore, a collapse of the algal community is not predicted.

Part B: Linkage Diagram

Because the stressor is a toxicant, it is expected that a structural response will occur first, in this case, growth inhibition of all copper-sensitive algal species and the extinction of the most sensitive species. A functional response will follow, with a reduction in the primary production of the algal community, plus an imbalance in P/R ratios, and export of unused primary production because of toxicity to herbivores. The structural response will change the algal community composition and reduce abundance, producing a habitat with a reduced algal food supply for herbivores. The remaining tolerant algae may also be less palatable to herbivores.

Example 2A:
Stressors: [Cu] = 0; [P] = Low
Baseline Conditions: undisturbed, oligotrophic environment

Part A: Stressor-Response Curve

The model predicts that when low concentrations of a limiting nutrient such as phosphorus are added to an oligotrophic system, a stimulatory response in the subsidy portion of the stressor-response curve will occur. Because baseline conditions are oligotrophic, the range of response will remain within the subsidy portion.

Part B: Linkage Diagram

The model predicts that there will be a rapid functional response of an oligotrophic system to low concentrations of phosphorus, measured as an increase in primary production. It is expected that the P/R ratio will remain balanced, because phosphorus concentrations will not be sufficient to push the oligotrophic system into an unbalanced, culturally eutrophic system. Because the baseline condition is oligotrophic, it is not expected that the increased phosphorus concentrations will produce a detectable change in algal community structure, because there will be insufficient change in overall nutrient status to produce a detectable shift in community composition (although biomass will increase). The functional response will cause a habitat change because food supply to herbivores will increase.

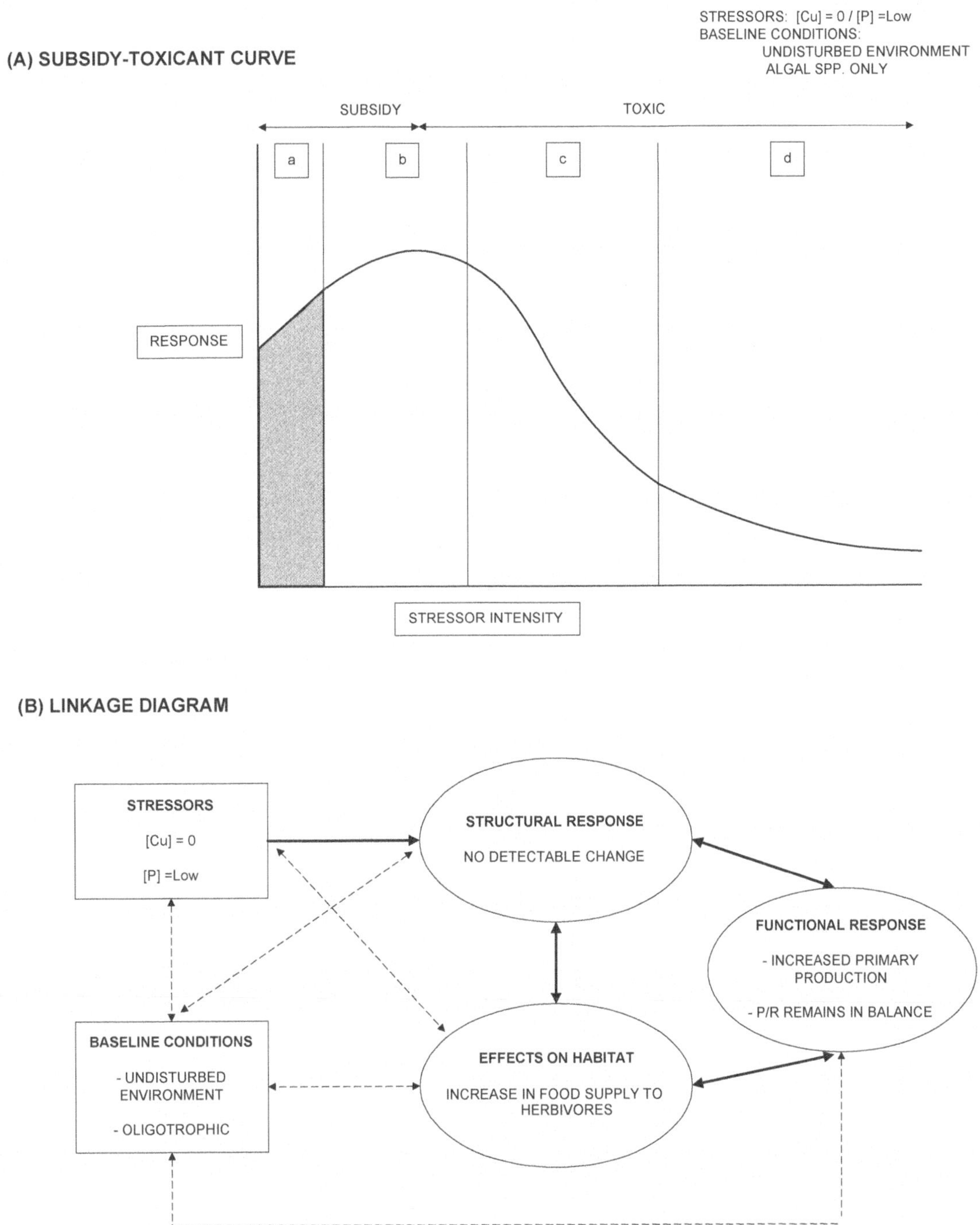

Figure 3-8. Conceptual Model Example 2A.

(A) SUBSIDY-TOXICANT CURVE

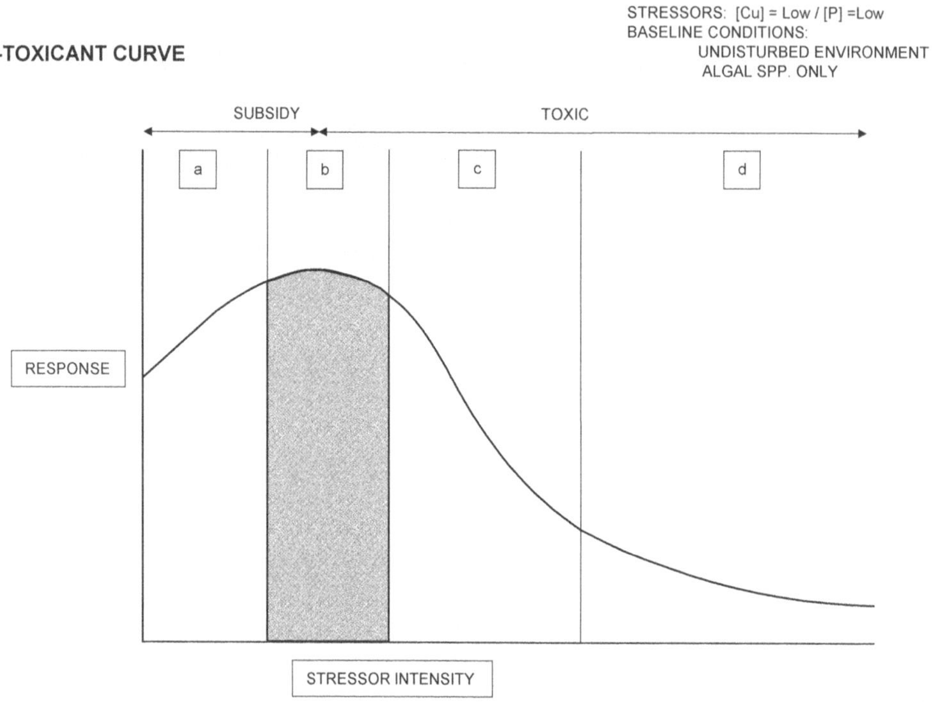

(B) LINKAGE DIAGRAM

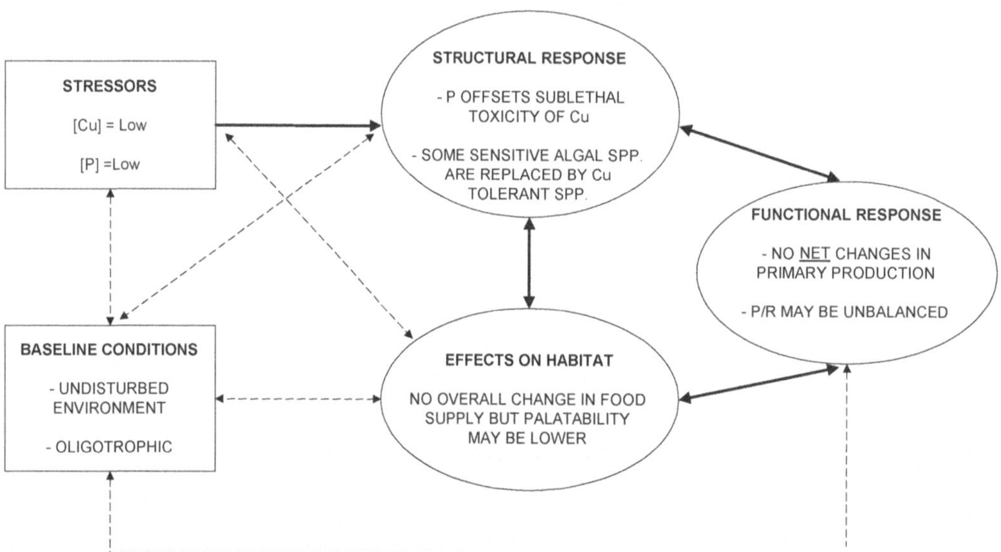

Figure 3-9. Conceptual Model Example 2B.

Example 2B:
Stressors: [Cu] = Low; [P] = Low
Baseline Conditions: undisturbed, oligotrophic environment

Part A: Stressor-Response Curve

In this example, the presence of a toxicant, copper, in low concentrations tends to offset the potential subsidy effect of low phosphorus concentrations. Responses are predicted to fall on both sides of the cusp between subsidy and toxicant, as illustrated by the shadowed portion under the curve.

Part B: Linkage Diagram

It is very difficult to predict whether a functional response to the phosphorus subsidy will occur before a structural response to toxic concentrations of copper. The overall, longer-term functional response in this case would be hardly noticeable because sensitive algae species will be replaced by copper-tolerant species with no net change in the primary production of the algal species. The P/R ratio may be <1 because of the energy costs associated with above-baseline copper concentrations. The overall structural response is predicted to be a shift in algal community structure to one dominated by copper-tolerant species. The net effect on habitat may be quite subtle, with no overall change in food abundance, but with a potential decrease in palatability.

Example 2C:
Stressors: [Cu] = High; [P] = Low
Baseline Conditions: undisturbed, oligotrophic environment

Part A: Stressor-Response Curve

The addition of copper in high concentrations would normally cause a pronounced effect on structure of the system. However, the addition of the nutrient phosphorus in low concentrations is expected partially to mitigate the deleterious effects of the toxicant. This places the predicted response in the toxicant region of the stressor-response curve; however, not to the point where the system completely collapses.

Part B: Linkage Diagram

Because the response is in the toxicant region of the curve, the model predicts that a structural change in the algal community will occur first. Most algal species would be inhibited and sensitive species would be eliminated. There may be blooms of copper-tolerant species because of the additional phosphorus in the system. The functional response would be reduced primary production reflecting the reduction or elimination of most algal species. The P/R ratio would be unbalanced (likely <1). There would be an export of unused primary production because copper toxicity would reduce the abundance of herbivores. Effects on habitat would include an alteration in food supply because of the reduction or elimination of sensitive algal species. The remaining tolerant species may be less palatable to herbivores. There may be low oxygen conditions caused by the decay of blooms of copper-tolerant algae.

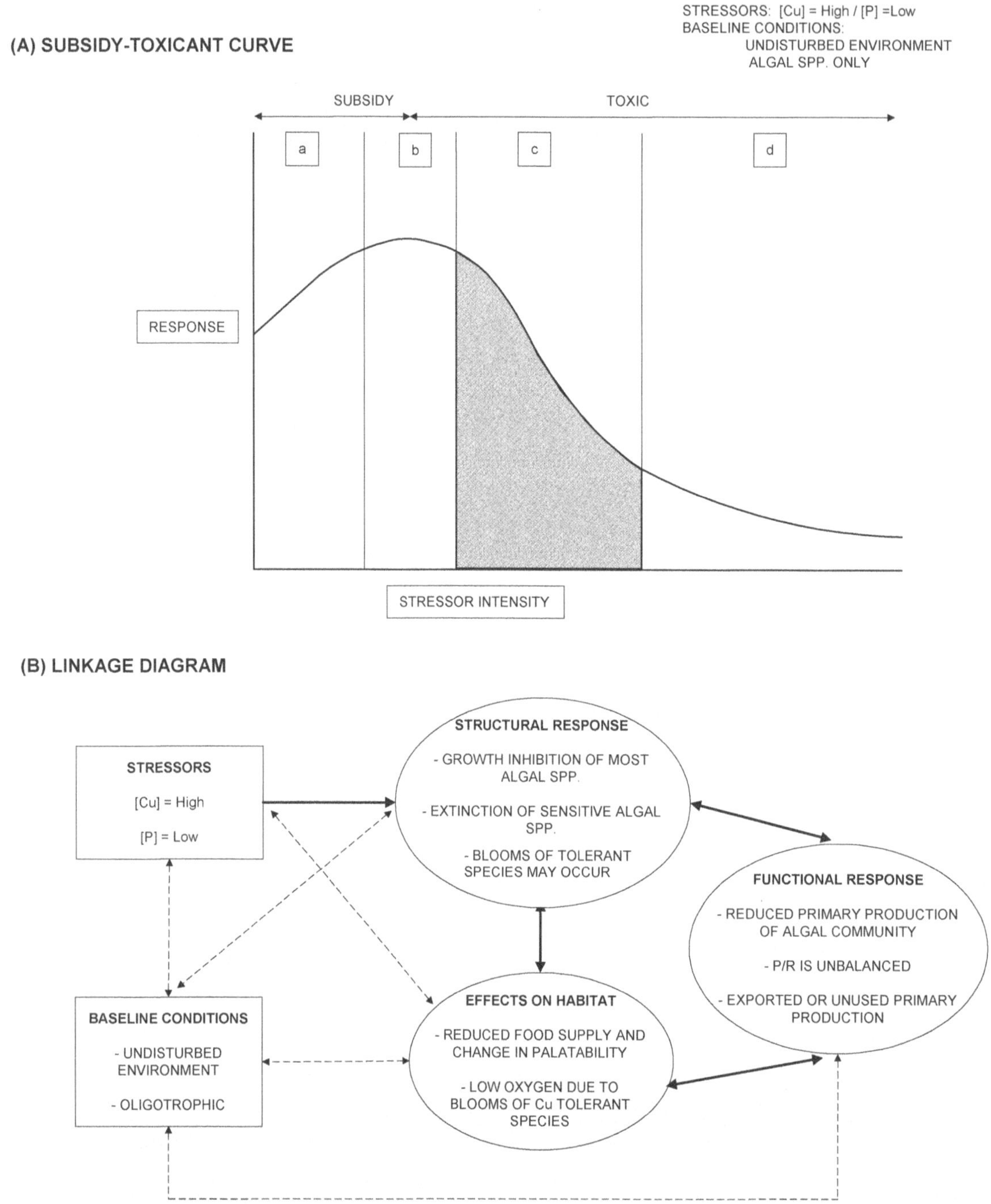

Figure 3-10. Conceptual Model Example 2C.

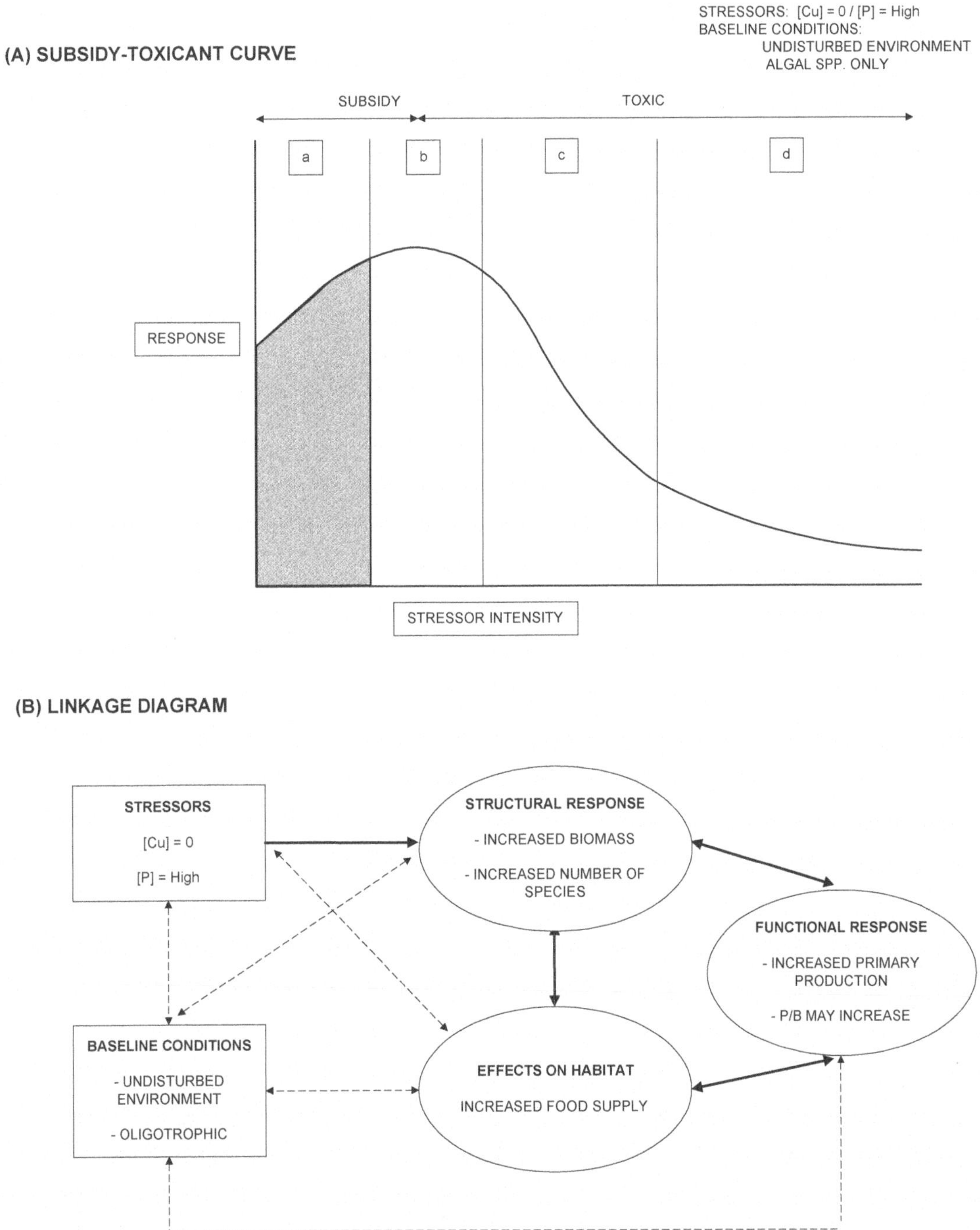

Figure 3-11. Conceptual Model Example 3A.

Example 3A:
Stressors: [Cu] = 0; [P] = High
Baseline Conditions: undisturbed, oligotrophic environment

Part A: Stressor-Response Curve

The addition of "high" concentrations of phosphorus to an oligotrophic system in the absence of copper is predicted to produce a subsidy response. It is not expected that this subsidy response will approach the "cusp" where subsidy converts to toxicant because of the initial oligotrophic condition. The ecosystem may shift to a mesotrophic status, but the deleterious effects associated with cultural eutrophication are not predicted.

Part B: Linkage Diagram:

A functional response is expected to occur first because the stressor falls within the subsidy range of the stressor-response curve. The functional responses would include increased primary production. The P/B ratio may increase if the enrichment shifts the community to smaller species. The structural response would include an increased number of species because of the increased availability of nutrients. A higher biomass would also be expected. The increased productivity and biomass would increase the food supply.

Example 3B:
Stressors: [Cu] = Low; [P] = High
Baseline Conditions: undisturbed environment

Part A: Stressor-Response Curve

This combination of stressors produces the widest range of possible responses. The actual response will depend upon the balance between the toxicant effect of copper and the subsidy effect of phosphorus. If the copper concentrations were low enough to still produce a hormetic effect in some algal species, then the combined subsidy effect with phosphorus would produce a response that falls solely in the subsidy range of the stressor-response curve. If copper concentrations exceed the tolerance threshold of several species, there may be a balance where the response oscillates between subsidy and toxicant regions, depending upon the ratio of sensitive/tolerant species. If copper concentrations exceed the tolerance threshold of the majority of algae species, then the response would be solely in the toxicant range of the curve.

Part B: Linkage Diagram

It is very difficult to predict whether structural or functional responses would occur first in this example because of the wide range of possible positions of the response on the subsidy-toxicant curve. Structural responses would include inhibition or elimination of copper-sensitive species, with replacement by tolerant species. The tolerant species may do very well because of the additional phosphorus, leading to an increase in primary productivity produced by fewer species. However, the presence of copper may increase energy requirements, pushing the P/R ratio to much less than one. Furthermore, there may be an export of unused primary productivity because of the reduction in copper-sensitive herbivores. Effects on habitat are equally difficult to predict. There may be no change in food supply, but palatability may decrease. Oxygen levels may decline because of the decay of blooms of copper-tolerant algae.

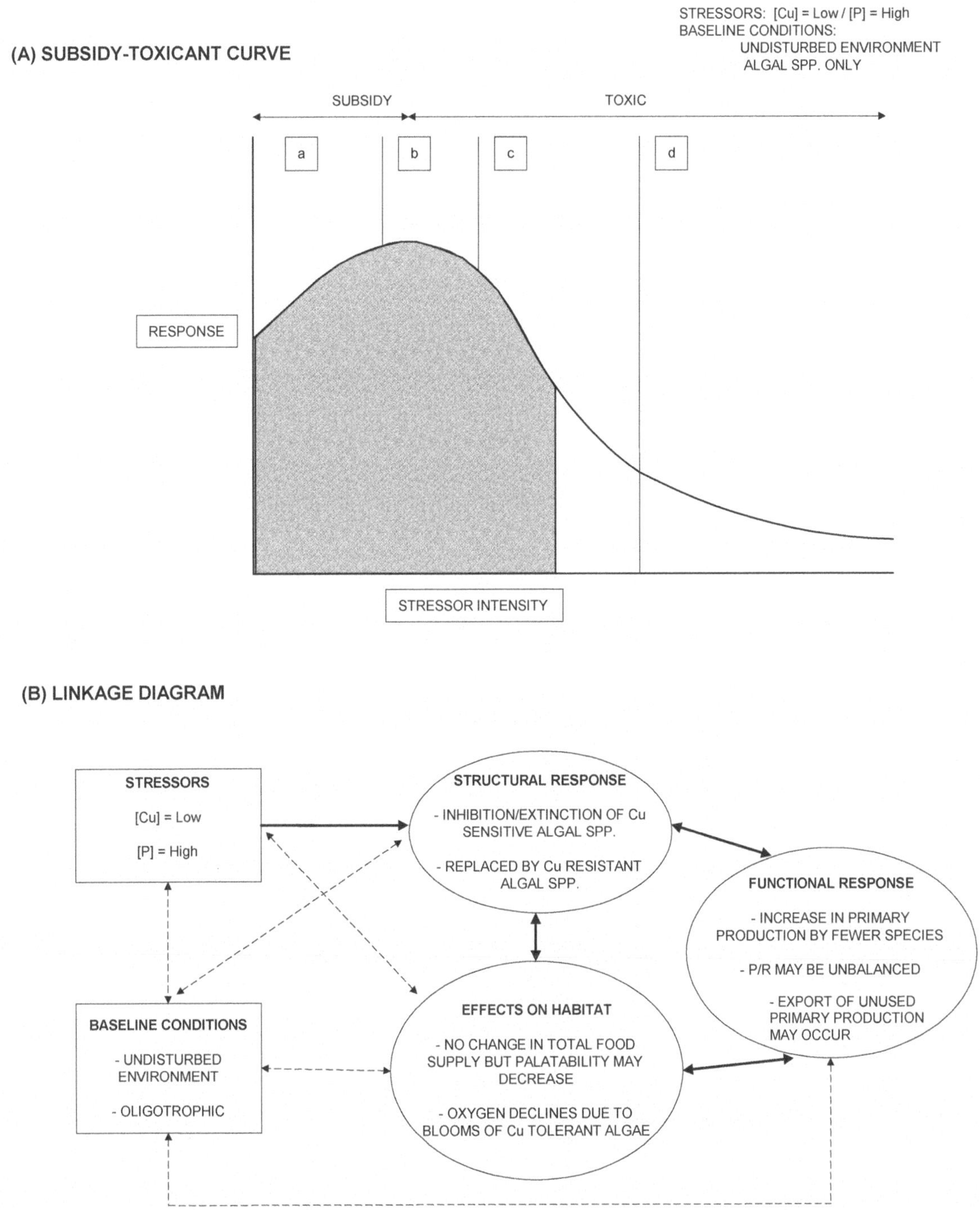

Figure 3-12. Conceptual Model Example 3B.

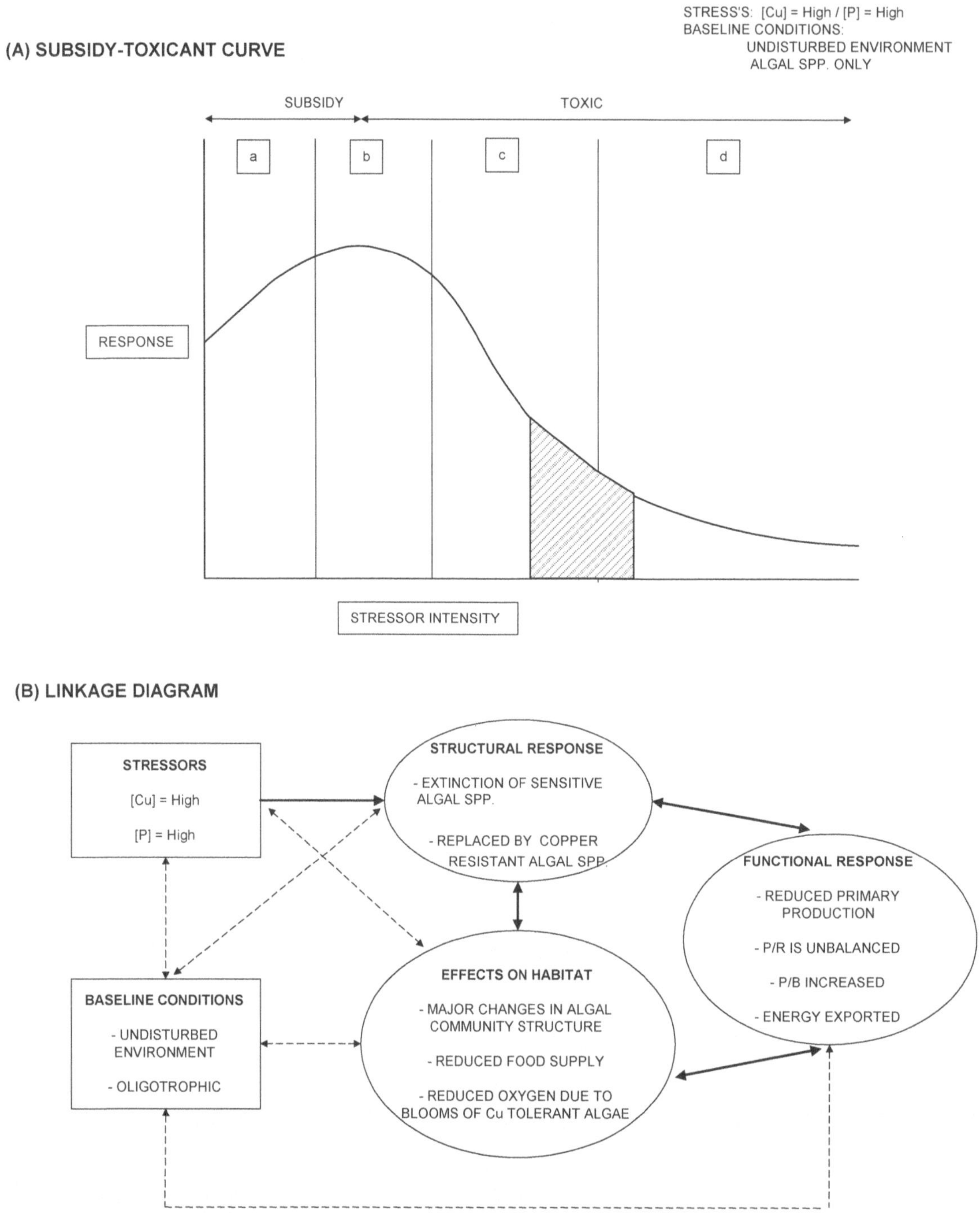

Figure 3-13. Conceptual Model Example 3C.

Example 3C:
Stressors: [Cu] = High; [P] = High
Baseline Conditions: undisturbed environment

Part A: Stressor-Response Curve

High copper and high phosphorus added to an oligotrophic system is predicted to produce responses in the toxicant zone of the stressor-response curve. The response may fall into the region reflecting severe change if the majority of algal species are copper-sensitive. In this example, the toxicant effects of high copper levels are expected to dominate. The subsidy effects of high phosphorus would only be observed in copper-tolerant species.

Part B: Linkage Diagram

The predicted response in the toxicant range of the subsidy-toxicant curve will produce a structural change first. Extinction of sensitive algal species will occur. Copper-resistant algal species will remain. These remaining species will have ample phosphorus; therefore, there will not be a total collapse of primary production. However, an overall reduction in primary production is predicted. The energy costs associated with high copper concentrations will produce P/R ratios <1. It is also expected that P/B ratios will increase. There will be energy export from the system as unused primary production is lost because of the reduction in copper-sensitive herbivores. Habitat effects will include a reduction in food supply. Oxygen sags may occur because of the decay of acid-tolerant algal blooms.

3.7 Tests of the Conceptual Model Under Differing Baseline Conditions

All examples presented above refer to the same initial baseline condition (i.e., an undisturbed, oligotrophic system). If baseline conditions were different, with toxicant concentrations above baseline and/or with a more enriched nutrient status, the *y* axis of the stressor-response curve would be shifted (as illustrated by the [b], [c], and [d] positions on the general stressor-response curve). This shift would produce different predicted outcomes from the above examples.

In a real-world situation, the baseline conditions must be factored into the formulation of the specific risk hypotheses to be tested. For example, if the baseline shifts the *y* axis to the [c] position, then Risk Hypothesis 1 (regarding functional responses occurring first when a subsidy response is expected) is not relevant. It is assumed that repeated testing of the conceptual model under varying baseline conditions will contribute to the understanding of the complex interactions between nutrient limitation (N or P), organic enrichment, the form of the potential toxicant (e.g., metal-ligand complexation), and the ecosystem response. For example, in most aquatic systems, additional total phosphorus concentrations will result in significant shifts in the ecosystem response. However, some systems may quickly become nitrogen-limited. The nitrogen limitation then produces a situation where additional phosphorus does not necessarily drive the ecosystem response to eutrophic or hypereutrophic conditions.

Baseline conditions will not be static; therefore, the conceptual model should be tested within the context of variable baseline conditions with season and year-to-year and results interpreted appropriately. For example, nitrogen limitation may occur in late summer in what are otherwise phosphorus-limited systems. Thus, the prediction of multiple-stressor effects on such a system would require specification of the season and corresponding baseline conditions.

The testing of the conceptual model will also require consideration of the theory of community conditioning. This theory suggests that ecological communities tend to preserve information about every event in their history and they are not at equilibrium, but rather are in the process of responding to their own unique history of interactions (Matthews et al., 1996; Landis et al., 2000; Sandberg and Landis, 2001). This was demonstrated by experiments where functional parameters of systems treated with a herbicide were not different statistically but the systems, on re-exposure, were much more resistant to the herbicide (Kersting and Van Wijngaarden, 1999). Thus, responses to multiple stressors may produce a state of the ecosystem that is different in each case. The resulting state may have equivalent functions to the original baseline and it may have acceptable community structure, but it will not be the same from one case to another, no matter how similar the original baselines may appear to be. This would pose a daunting challenge to the objective of producing reliable predictions of the response to multiple stressors.

The goal of this study is to provide a science-based study design that helps establish a general, conceptual approach to multiple stressors. Therefore, testing of the applicability of the conceptual model across a range of baseline conditions with varying histories (and presumably different types of "conditioning") will be required. A comprehensive set of tests under different baseline conditions would follow after initial investigations have shown whether the risk hypotheses for the conceptual model are proven or disproved.

3.8 Role in Pollution Control Decision-Making

The science-based approach of this model constitutes an important contribution to the unresolved question of cause and effect relationships when multiple stressors are discharged into an aquatic environment. It is a tool that can significantly contribute to a more cost-effective approach in decision-making by helping the pollution control authorities target their spending efforts to the stressor(s) of concern that pose the most risk under multiple-stressor conditions. The model also allows interpretation of risk under different baseline conditions.

The outputs of this study that are relevant for decision makers are:

- Ranking of the stressor or combination of stressors for particular baseline conditions.
- Determination of threshold concentrations for a particular stressor or combination of stressors, for a known baseline.

Table 3-3 summarizes how the model's outputs can contribute to the decision-making process. The remedial measures suggested here are intended for point source contributions alone. A "mass-balance" of a particular stressor(s) must be performed before any decisions are made at a WWTP level. For example, if a stressor(s) is a nutrient (particularly phosphorus), nonpoint source contributions can be just as important as or even more so than point source contributions.

The model's outputs are powerful tools that may be used in a "global" perspective (i.e., the model can allow predictions to be used confidently in diverse locations because it can be used to determine the threshold concentration of a stressor or combination of stressors for different baseline conditions). This ability allows the prediction of the assimilative capacity of a particular ecosystem, by determining the maximum stressor(s) concentration/intensity that can be withstood before ecosystem integrity is adversely affected.

Table 3-3. Examples of Possible Effects of Conceptual Modelís Outputs in the Wastewater Treatment Industryís Decision-Making Process (for Point Source Contributions Only).

Are Stressor(s) Concentrations Deleterious to Ecosystem?	Current WWTP status in relation to stressor removal	Possible Control/ Remediation Measures
Stressor combination produces responses in the subsidy portion of the curve **(Subsidy Stage)**	WWTP already removes stressor(s)	No further action may be required **OR**
		Deactivate treatment units/systems when stressor concentrations are less than stressor threshold concentrations
		Reduce amount of agents (chemicals etc.) used for stressor removal, as appropriate
		Combination of measures
		Other site-specific Measures
Stressor combination produces responses in the subsidy portion of the curve **(Subsidy Stage)**	WWTP does not remove stressor(s)	Check forecasted changes in effluent quality
		Determine when specific treatment upgrades or additions will be needed
		Consider source control of stressor(s)
		Combination of measures
		Other site specific measures
Stressor combination produces responses in the toxicant portion of the curve; stressor(s) contributing the most to toxic response identified **(Toxic Stage)**	WWTP already removes stressor(s)contributing the most to toxic response **Moderate measures required**	Upgrade the WWTS to remove the stressor(s) to acceptable levels, i.e. stressor concentrations less than stressor threshold concentrations
		Expand the WWTS to further remove the stressor(s) to acceptable levels, i.e., stressor concentrations less than stressor threshold concentration
		Consider source control of stressor(s)
		Combination of strategies
		Other site-specific measures
Stressor concentrations are the same as above **(Toxic Stage)**	WWTP does not remove the stressor(s) contributing the most to toxic response **Urgent measures required**	Secure funds to reduce stressor concentrations to less than stressor threshold concentrations by one or a combination of:
		Expand the WWTP. Add treatment module/agent/ other for stressor removal
		Upgrade the WWTP to remove the stressor(s) so stressor concentrations to less than stressor threshold concentrations
		Consider control of stressor(s) at their source
		Combination of strategies
		Other site-specific measures

CHAPTER 4.0

EXPERIMENTAL DESIGN

4.1 Key Research Questions

The key question underlying this project is: How do lotic ecosystems respond to multiple stressors? The proposed conceptual model is a means of addressing that question. The experimental testing of the model will determine the robustness and applicability of the proposed conceptual model.

The three risk hypotheses in the conceptual model are: 1) subsidy stressors first cause a functional response; 2) toxicant stressors first cause a structural response; and 3) multiple stressors will act in an additive manner.

To test these risk hypotheses, we must ask:

- Do subsidies consistently cause a functional response in advance of structural changes?
- Do toxicants consistently cause structural changes in advance of significant functional changes?
- Are there definable cusps (thresholds) at which the influence of a stressor switches from a subsidy to a toxicant?

Also, there is a related question that must be answered: Are the effects of subsidies or toxicants different or more profound when they affect a keystone species? A keystone species plays a role affecting many other organisms in an ecosystem. The loss of a keystone species can lead to sharp population drops and extinction of other species that depend on it for certain services such as nutrient cycling, primary food supply, or provision of habitat structure (e.g., macrophytes provide cover). Therefore, it is very important to understand whether effects of multiple stressors on keystone species will overwhelm the compensatory mechanisms and redundancy built into aquatic ecosystems.

Answering these questions is necessary to validate the conceptual model. When we have confirmed, or modified, the model, we can then turn to a central question: Is the effect of multiple subsidies or stressors additive?

This is a simple question, but answering it is rather more complex. To begin, we must define what we mean by additive. Stressors like heat and copper cannot be quantitatively compared, but we can compare their effects on a common scale. For example, if a given concentration of copper causes a 50% reduction in algal production compared with baseline conditions, and a temperature increase of 5°C augments production by 20%, we would expect that adding copper at that concentration in water warmed by 5°C would reduce algal production by 30%. Assuming our production estimates were sufficiently accurate, a change in algal production that is not significantly different than a 30% reduction would be taken as evidence of additivity. In practice, the determination of additivity would be made by the significance of interaction terms in an ANOVA, as described in Section 4.10.

The above example is deliberately simplistic. We know, for example, that temperature changes cause a nonlinear change in metabolic rates, therefore, the toxicity of copper does not necessarily change linearly with temperature. In addition, the rates of various chemical reactions determining the form and availability of copper are also temperature dependent. However, our model is intended to apply, as Odum's model on which it is founded, to responses at the community- and whole-ecosystem levels. When the toxicity of copper and the stimulatory effects of higher temperature are integrated across entire communities of epiphytic algae, an additive response is not an unreasonable null hypothesis. When we consider the prospects for indirect effects, for instance through augmentation (from higher temperature) or depression (from copper toxicity) of invertebrate grazing, it becomes evident that the known effects of temperature and copper on metabolism of individual organisms cannot be extrapolated to higher levels of organization.

There are two alternatives to additivity of stressors. The more obvious is that combined stressors disturb ecosystems more severely than the sum of their effects individually. To continue the earlier example, a synergism between copper and temperature could occur such that the increase in toxicity at higher temperature overwhelms any temperature-mediated acceleration of production. This situation could arise in an oligotrophic ecosystem in which plant production is limited by phosphorus and hence hardly responds to temperature. Alternatively, the stimulated metabolism in warmer water might permit the algae to resist copper more effectively, resulting in a lesser depression of algal production than predicted.

The latter effect, in which combinations of stressors are less potent than their individual effects would predict, is congruent with the conceptual model. Species within the community are not equal in their sensitivity to stressors, but there is evidence, among the invertebrates at least, that tolerance tends to be correlated. Species that are resistant to metals toxicity also tend to be resistant to temperature, or low DO, or a number of other stressors. The resistant species are better adapted to more severe environments generally. If stressors such as excess heat or copper both remove or depress the same species in the community, their combination may have relatively little additional effect because the susceptible species have already been removed. The resistant species remaining are unaffected by both stressors. In the model, this situation would be expressed as each stressor pushing the response curve to the right at position [d] into the zone where the curve flattens and there is relatively little additional response.

The discussion to this point has assumed that the effects of subsidies and toxicants will be identical alone as together. However, the subsidy/toxicant model predicts a more complicated interaction. Adding a toxicant to a system receiving a subsidy may not only reduce the subsidy effect, it may move the cusp between subsidy and toxicant to the left, that is, toward causing a deleterious effect at a lower concentration.

To adequately address the question of additivity, we need to examine the effects of several different kinds of toxicants and subsidies, at several different concentrations or intensities, on several different response endpoints. Only by combining information from these diverse measurements can we begin to assemble a case for or against additivity. In this round of experiments, it is essential to consider the three fundamental combinations of stressor: 1) toxicant plus toxicant; 2) subsidy plus subsidy; and 3) subsidy plus toxicant.

The combinations to be used and the level of each stressor must be carefully considered. One prediction from the subsidy/toxicant model is that the cusp between subsidy and toxicant will occur at a lower concentration (that is, the curve shifts to the left) in the presence of a second stressor. Depending on whether the effect of multiple stressors departs from additivity, a poorly chosen subsidy might become a toxicant when a second stressor is applied to the system. Hence, a subsidy like a small increase in temperature may become a toxicant when it is combined with another stressor like toxic levels of copper.

This possibility complicates tests of additivity, because we cannot be sure in such instances whether we are adding two subsidies or a subsidy and a toxicant. For now, it is best to avoid ambiguous situations by choosing combinations that are firmly in one camp or the other. Because the subsidy region of the concentration curve is small relative to the toxicant region (and for some stressors, may not exist), it is best to begin with combinations of toxicants. If those tests are successful, the more difficult studies combining subsidies could be undertaken. Only when these two options have been adequately explored should we attempt combinations of a subsidy and a toxicant. In summary, it is strongly recommended that the experiments test subsidies and stressors in the order listed above.

4.2 The Choice of Study Methods and Suggested General Test Scheme

The effect of stressors from point source and nonpoint source discharges in lotic systems has traditionally been evaluated using the results of standardized single-species bioassays combined with receiving environment monitoring data. Although this approach has been effective in providing feedback regarding the effectiveness of wastewater treatment systems and the level of protection provided to the receiving environment, the ability to extrapolate to other situations with a different set of multiple stressors is limited. In addition, the existing knowledge base acquired through standard toxicity testing and field observational studies has not produced a set of central theories regarding the effects of multiple stressors that have been tested sufficiently to allow confident prediction.

The goal of this project is to provide a study design that helps establish a general, conceptual approach to multiple stressors. The tools used to meet this goal must be capable of testing the conceptual model. Furthermore, the tools must provide useful lines of evidence within a weight-of-evidence approach, because the combined use of mesocosms, field surveys and laboratory studies has been shown to aid the development of convincing cause-and-effect arguments through the application of weight-of-evidence criteria (Culp et al., 2000a; Lowell et al., 2000). Methods commonly used to measure indicators or stressors can be grouped into several categories: laboratory studies, stream microcosm studies (simulated ecosystem studies including microcosm and mesocosm systems), and direct field biomonitoring.

Stream microcosms, also known as artificial streams, are simulated ecosystem experimental units that are used to examine many levels of biological organization, ranging from single-species tests at the individual organism level to multispecies tests at the community level. They vary widely in design, from simple laboratory systems (microcosms) to medium sized portable units (mesocosms) to elaborate outdoor complexes (large-scale mesocosms, also called "artificial streams") (Guckert, 1993).

For the purpose of this report, the three investigation tools will be referred to and can be described as:

- Laboratory studies—experimentation in the laboratory using single species at the individual organism level (e.g., toxicology testing, dose-response experiments on a selected algal species, etc.);
- Mesocosm studies—experimentation using medium-scale mesocosms investigating multispecies at the community level (e.g., dose-response experiments on various algal and/or benthic invertebrate species), also includes large-scale artificial stream mesocosms (usually large outdoor facilities);
- Field studies—field biomonitoring on relevant sites.

The General Test Scheme proposed would start with mesocosm experiments because formal hypothesis testing using controlled experiments is required to initiate the process of evaluating the applicability of the conceptual model. Applying this triad of laboratory, mesocosm, and field studies can help to establish plausible mechanisms and specificity of effects, dose-response curves, and assessments of the concordance between field and experimental results.

The study team suggests a combination of the three main study tools (laboratory tests, mesocosms, and field studies) in a General Test Scheme (Figure 4-1). Each of these tools has merits and drawbacks, and each can supply valuable information on certain aspects of stressors' effects. Field studies provide critical observational evidence, including spatial and temporal correlation between stressor exposure and effects, a high level of ecological realism, and information on bioaccumulation of contaminants. However, field studies cannot be used to formally test risk hypotheses. Laboratory tests can provide experimental verification of stressor effects and indicate the strength and specificity of biological responses to this stressor; however, it is difficult to directly relate these results to field scenarios. Furthermore, simple laboratory tests on their own may indicate a likelihood of stressor effects that may not occur in the field. This is because laboratory tests cannot completely simulate complex field conditions and biotic interactions.

Field-deployed mesocosms, of which there are many types (Lowell et al., 1995a; Dyer and Belanger 1999; Culp et al., 2000b; Culp et al., 2002; Dyer et al., 1998), provide a feasible and reasonably well-developed approach for bridging between laboratory experiments and field observations. Mesocosms allow testing of risk hypotheses under field environmental conditions, and can be designed for various purposes. Figure 4-2 illustrates two different types of mesocosms used.

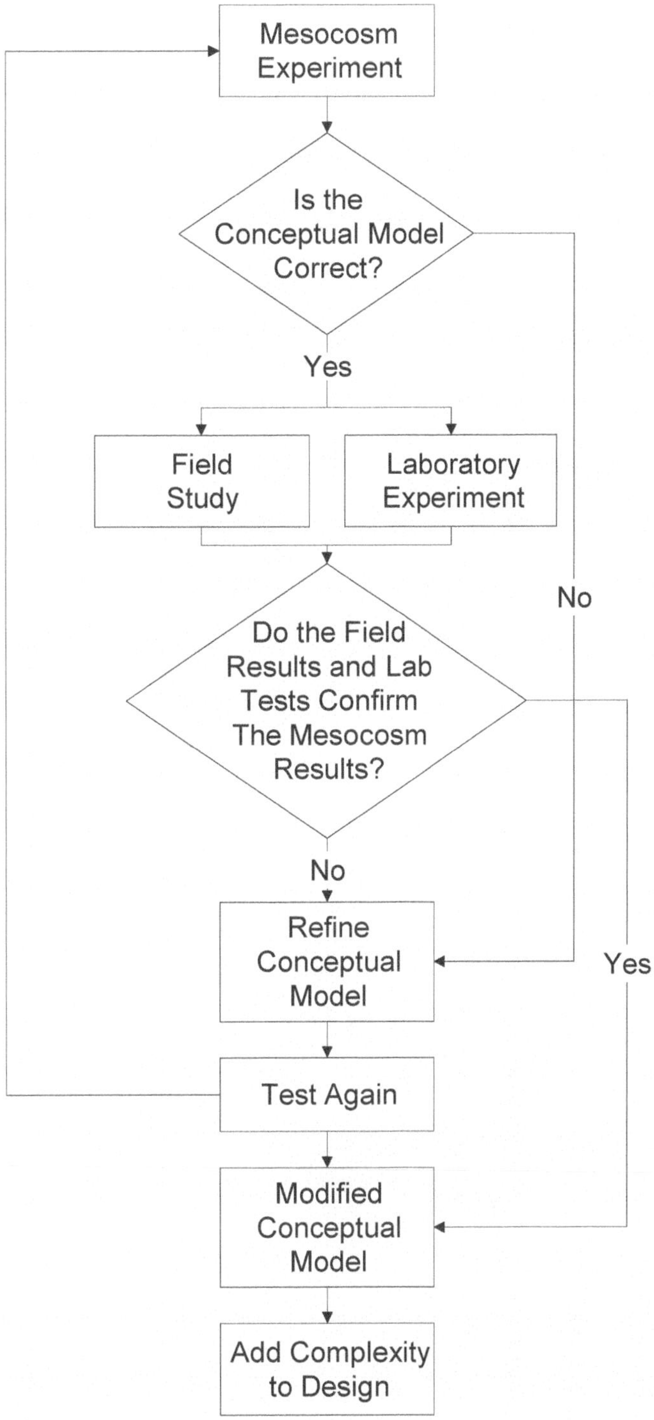

Figure 4-1. General Testing Scheme.

Although the study team is aware of the advantages and disadvantages of each of the three approaches (Table 4-1), we suggest that mesocosms be the key experimental tool within the General Test Scheme (Figure 4-1). The following discussion presents the reasons for selection of mesocosms as the primary tool within the General Test Scheme and the recommended use of laboratory and field studies within the overall weight-of-evidence approach. A detailed review of mesocosms designs and approaches is provided in the accompanying report, "Multiple Stressors: Literature Review and Gap Analysis" (Swanson et al., 2004; 00-ECO-2B).

A) Mobile mesocosm system with streams situated on two trailers (developed by Natural Water Research Institute of Environment Canada, Saskatoon, SK).

B) Mobile mesocosm system with streams situated on a single trailer developed by AMEC Earth and Environmental, Fredericton, NB).

Figure 4-2. Examples of Different Types of Mesocosm Designs Used for Environmental Effects Monitoring in Canadian Rivers (Reprinted with permission from DubÈ et al., 2002).

Table 4-1. Comparison of Laboratory, Mesocosm, and Field Studies.

Issue	Laboratory Studies	Mesocosm Studies	Field Stream Studies
Ecological relevance/realism	Not realistic. Usually static (batch reactors), occasionally flow-through (continuous flow reactors) Cannot accommodate more than a few confounding or correlated variables (e.g., water hardness, pH).	Can be placed outdoors near the study area, yielding more realistic exposure to environmental variables. Ecologically more relevant than laboratory studies but are a simplification of ecosystems. The results may not always indicate the changes in a ecosystem as a whole	Provides the most realism Allows comparison over wide spatial or temporal scales.
Cause/effect Relationship	Can be used to distinguish effects between stressors; good for screening purposes; Produce dose-response information.	Controlled nature of the experiments allows some cause/effect inference, particularly for testing of multiple-stressor effects	Cause/effect conclusions cannot be drawn from field studies, although inferences can be made and used as a basis for controlled experiments.
Effects thresholds	Limited to suborganism or whole organism measures.	Multilevel endpoints: individual, population & community level endpoints Can be used to investigate both structural and functional endpoints in populations and communities.	Multilevel endpoints: Individual-, population-, and community-level endpoints. Allow for simultaneous observation of several trophic levels
Stressor ranking	Useful for ranking of stressors for relative toxicity to single species	Useful for ranking of stressor combinations	Cannot be used for definitive stressor ranking, but weight of evidence can be applied
Direct vs. indirect effects	Can be used to observe direct effects of stressors on single species	Can accommodate both direct and indirect effects; unable to follow multigenerational effects.	Allow for observation of direct and indirect effects; able to follow multigenerational effects.
Controlled conditions	Variables can be strictly controlled	Relevant variables can be controlled, but less than in laboratory settings	Extremely hard to achieve
Replication	Easy replication	Replication more limited because of cost	A dilemma caused by confounding environmental parameters
Relevant test species	Single-species studies, often using standard test species.	Native species can be used, yielding more realistic results	Native species
Cost	Less expensive	More expensive than laboratory studies	May be more expensive than mesocosms, and will be case specific
Logistics	Relatively easy to set up	Requires access, provision for waste disposal; therefore, special permits	Site specific, may be extremely challenging

Issue	Laboratory Studies	Mesocosm Studies	Field Stream Studies
Confounding variables	Cannot account for all the confounding variables present in the field	Mesocosms can be used to examine the role of confounding variables, but the number of variable combinations will be limited by the mesocosm design.	Role of confounding variables examined by statistical inference. Better as a validation tool for mesocosm observations Provides measures of the result of complex ecosystem-level interactions
Multispecies tests	Effective for single-species testing at organism level	Allows for multiple species testing at the community level	Examines *in situ* multispecies communities.
Ease of use	Are simple and small scale	Needs more care and maintenance than laboratory reactors because of scale and complexity	Optimal design method has to be applied to reduce cost and labor
Repeatability	Low cost and simplicity allow for repeatability	Only a few runs are usually feasible because of cost limitations	N/A
Scalability (temporal/spatial)	No time scale assessments can be made. Scaling up to field-relevant spatial scales difficult-to-impossible.	Long-term studies possible but more challenging. Spatial scaling of mesocosms has to be done with care and precision	Excessive spatial heterogeneity can confound the interpretation of spatial trends. Temporal trends can be examined with careful design.
Fate/transport	Specific fate parameters can be investigated (e.g., K_{ow}, K_d, can be determined).	Partitioning among water, sediments, and biota can be measured but temporal/spatial scale limitations may produce invalid results.	Partitioning among water, sediment, and biota can be measured and rates of transport and deposition can be determined
Statistical rigor	Number of replicates and repetitions allow for statistical data treatment	Statistical analyses are relatively straightforward. Allows for treatment replication and testing hypothesis. May give false positive and false negative results caused by secondary effects, if these effects are not controlled.	May give false positive or false negative results caused by secondary effects in the field and inadequate statistical power.
Interactions among stressors	Better for screening individual stressors than multiple stressors	Can be used to study interactions between different stressors, including metals, herbicides, surfactants, nutrients, siltation, temperature, etc.	Can be used to assess biological interactions (e.g., predation) but only by statistical inference.

Compared to single-species studies in the laboratory, community-level studies typical of mesocosm experiments incorporate greater environmental realism, are better able to identify components within the community most sensitive to a given stressor, and integrate a variety of ecological processes (e.g., competition and food web dynamics) (Forbes and Forbes, 1994). Flowing-water mesocosm techniques can be used to identify mechanisms of stressor effects on riverine food webs while eliminating confounding environmental factors that are common to field observational studies (Culp et al., 2000a). Mesocosm technologies provide the means for experimentally manipulating complex riverine communities under ambient physiochemical field conditions of substrate, current, irradiance, water temperature, and water quality (Lamberti and Steinman, 1993). The application of mesocosm approaches also allows community- and ecosystem-level questions to be posed, and chronic exposures applied, at a more manageable scale than that of whole ecosystems (Kimball and Levin, 1985).

Increased community-level accuracy provided by mesocosms comes at the expense of decreased precision (more uncontrolled variables) relative to single-species studies and decreased ability to resolve some of the specific mechanisms responsible for observed responses (Culp et al., 2000b). Nevertheless, with specific objectives stated in advance of testing, mesocosm results can be related laboratory tests to help identify the effects of chemical and effluent stressors on communities. Indeed, the linkage of mesocosm and laboratory tests can be a powerful tool that provides a means of adding important ecological information to risk assessments (Shaw and Kennedy, 1996).

The limited ability to predict the effects of stressors in the field using only laboratory studies has been well documented (Dickson et al., 1996; Clements and Kiffney, 1996). For example, laboratory tests may be predictive of field responses when stressor effects are extreme and fairly obvious (e.g., acute mortality). However, when ambient toxicity is low (the more common scenario), directly relating laboratory results to field responses will be more difficult (Dickson et al., 1992; Clements and Kiffney, 1994), and more environmentally realistic approaches such as mesocosm experiments are appropriate.

Although laboratory studies are not the primary tool in the General Test Scheme, they are useful within an overall weight-of-evidence approach. Laboratory studies can be used to determine benchmark values for use in mesocosm studies in addition to providing stand-alone evidence of causal mechanisms to contribute to the conclusions reached using a weight-of-evidence approach. One of the roles for laboratory experiments is to conduct preliminary, standard toxicity tests that provide direct cause and effect information for a given stressor. Laboratory studies also allow for the direct investigation of functional endpoints and can provide evidence of response mechanisms.

Another important goal of laboratory experiments is to determine the appropriate dosing regimes in mesocosm experiments with tests specific to major groups present in the mesocosm assemblages (e.g., periphyton, benthic invertebrates, and fish). As discussed in Section 4.4, laboratory studies provide sufficient replication for the use of a regression approach so that the shape of the subsidy and/or toxicant response curve can be determined. Laboratory testing would, thus, lead to determinations of the most appropriate subsidy/toxicant concentrations and to facilitate distinguishing between subsidy and toxicant responses.

Field (*in situ*) stream investigations provide environmental realism at the community and ecosystem level (Table 4-1). Field data can provide evidence of correlations and interactions among variables. Field investigations provide the ecosystem perspective. They help shift focus from small- to large-scale phenomena, from population to community, and from structure to process (Rosenberg and Resh, 1993). Field data can help validate predictions made using mesocosm results. However, the environmental realism of field data is obtained at the expense of being able to control variables and formally test hypotheses. Field studies can be expensive and difficult. Data interpretation can be very challenging. It is often not clear whether field data can be applied to sites or times with environmental variables lying outside the range observed during the study. Conclusions based on comparisons with reference sites can be spurious if the reference sites have too many differences in baseline physical, chemical, and biological variables from exposed sites. Regression or multivariate analyses of field data can give inaccurate results if the variables are highly correlated.

Under an integrated General Test Scheme (Figure 4-1) concordance of results between laboratory studies, field observations, and dose-response exposures in mesocosms can provide the basis for a weight of evidence that can be used to refine the conceptual model and provide a robust predictive model for use by risk managers (Culp et al., 2000b). Notwithstanding the debates over "simple versus complex, field versus laboratory, and reductionist versus holist" (Clements and Kiffney, 1996), it is clear that laboratory tests and field studies are and can be an important part of a weight-of-evidence approach to the assessment of multiple stressors and the testing of the conceptual model.

The general scheme in Figure 4-1 can be used as testing of the risk hypotheses progresses through the recommended stressor combinations (i.e., toxicant plus toxicant, then subsidy plus subsidy, and finally toxicant plus subsidy). The conceptual model will continuously be refined with each iteration of the general testing scheme. Additions to the conceptual model to account for the results of each iteration are fully expected; however, the overriding goal will be to identify central, common tendencies. The conceptual model will only be as useful as its ability to correctly predict responses to multiple stressors under a variety of conditions.

Results obtained by laboratory tests and mesocosm studies should be complemented by observational evidence obtained *in situ*, via field investigations, as these provide temporal and spatial observational data (e.g., along a river gradient, or data from upstream and downstream locations from a point source) for selected ecological endpoints.

The combined data from laboratory testing, riverside mesocosm experiments, and field observations can be integrated in a weight of evidence that provides logical guidelines for establishing causation in ecological risk assessment. Example 4-1 shows this process and is extracted from the findings reported for the Fraser River Study (Culp et al., 2000a).

Example 4-1 illustrates how the weight-of-evidence approach makes use of multiple sources of information to develop scientific support for each criterion (refer to Culp et al., 2000a). For example, toxicity tests in the laboratory can advance the mechanistic understanding of stressor effects on the food web. This approach directly relates routine and novel laboratory toxicity test results to field experiments and surveys.

Example 4-1. Pulp Mill Effluent.

a. Laboratory Testing:
Included nutrient limitation studies on bacterial production and toxicity tests that investigated the effects of pulp mill effluent (PME) on algal growth, invertebrate survival, and reproduction and early life stage survival of fish.

Results indicated that:
Phosphorus (P) limited development of bacteria, similar to findings using 3% PME.
Toxicity was only an issue at high PME concentrations
Low PME concentrations showed growth stimulation for algae and increased production for the invertebrate.

b. Mesocosm Experiments:
Community-level studies using three treatments that simulated PME concentrations in the Fraser River, B.C., at different times and locations.

Investigated the effects that each treatment had on periphyton and benthic invertebrate biomass as well as in structural composition, using several effects measurements. Also investigated toxicity of PME to benthic invertebrates, as well as bioaccumulation of various contaminants in selected periphyton endpoints.

Results indicated that the primary effect of exposure to the various treatments is increased river productivity (as measured via periphyton biomass and dominant insect taxa) via nutrient enrichment rather than toxicity. It also pointed to bioaccumulation of chlorinated organic compounds in algae for higher PME exposures.

c. Field Surveys:
The field investigations included determination of the following measurements in the river, upstream, and downstream of the discharge: algal and benthic invertebrate biomass, algal species composition, and existing data on concentrations of various contaminants in sediments and benthic biota, collected downstream of six upper Fraser River Basin pulp mills.

The results indicated an increase in algal and benthic invertebrate biomass downstream of the discharge, with bioaccumulation of chlorinated compounds in insects and fish.

Overall Analysis

Data from the three sources supported a number of common causal criteria, described in detail in the article. Data from the field surveys confirmed the findings obtained by the mesocosm studies, for all treatments investigated, and laboratory toxicity testing provided evidence that corroborated field mesocosm results for primary and secondary producers, namely that dilute concentrations of PME stimulate production within the benthic food web. Using a weight-of-evidence approach, it was found that field surveys corroborated the findings from the mesocosm study, in that there is an increase in river productivity downstream of the discharge via enrichment (Nutrient → periphyton → insect grazers). It also confirmed that the discharge did not cause any shift in the algal dominant species, diatoms.

As previously mentioned, the integrated use of field, laboratory, and stream mesocosm data using weight-of-evidence causal criteria constitutes a powerful tool for analyzing ecosystem responses to multiple stressors. When this approach is used in investigations conducted at different sites (and conditions), it is possible to develop and test plausible conceptual models that describe the ecosystems' response to the tested stressors under a variety of conditions.

Figure 4-3 illustrates a conceptual model that was generated in this manner (i.e., by integrating field, laboratory and stream mesocosm data using weight-of-evidence causal criteria). The data were obtained from several investigations including the Fraser River study, briefly described in example 4-1 (Culp et al., 2000a), and from the Thompson River and Athabasca River studies (Culp et al., 2000a, citing others).

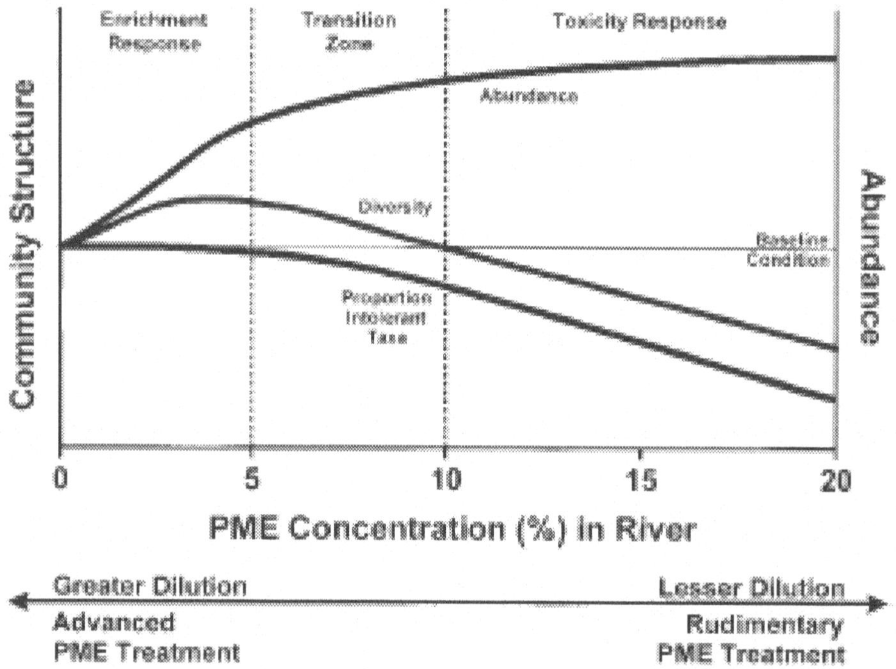

Hypothetical response curves for benthic community abundance and structure, and the relative importance of pollution-intolerant taxa as pulp mill effluent (PME) concentration in large oligotrophic rivers increases. The baseline represents expected relative values for reference river reaches. Note that PME concentration varies along the abscissa as a result of decreasing dilution related to longitudinal mixing and seasonal changes in the hydrograph, or because of modifications in effluent treatment processing (Culp et al., 2000a).

Figure 4-3. Conceptual Model Describing Benthic Community Structure Change Resulting From Exposure to Pulp Mill Effluent in Large Oligotrophic Rivers. Reprinted from Culp et al., *Environ. Toxicol. Chem.* 2000, 19, 1167-1173. Copyright 2000, SETAC, Pensacola, Florida, USA.

4.3 Effects Endpoints

Effects endpoints best suited for study in mesocosms with complementary data obtained from laboratory studies and field investigations are those that can be used regardless of the scale of the experiment or the size of the field study area; have been shown to respond in a predictable manner to stressor gradients; are sensitive indicators of either functional or structural change; are proven and well accepted; and are cost effective and practical.

Some of the endpoints that meet these criteria are presented in Table 4-2, which illustrates how laboratory, mesocosm, and field studies can be used to complement each other. Effects endpoints that can be used in all three types of studies are most useful when building a weight of evidence for cause/effect relationships among multiple stressors and effects.

Effects endpoints such as biomass (wet and dry weight), abundance (cell counts, number of individuals/m^2), and chlorophyll *a* concentration can be used in all three study types (Table 4-2). These endpoints are indicators of functional response and abundance can also indicate structural response. All three can be expressed in units that are proportional to some standard area or volume. Confounding variables that may bias the results (such as experimentally related light limitation) can be accounted for. The response of these endpoints to a range of stressors has been well-documented in the literature. Biomass, abundance, and chlorophyll *a* have been used by scientists for decades and are well accepted, cost effective, and practical (Edmondson and Winberg, 1971). If responses of these effects endpoints in laboratory, mesocosm, and field studies are consistent and coherent, and show a strong correlation with the stressor concentrations, then a convincing weight of evidence can be presented.

Other, less routine, functional metrics include carbon-14 uptake (as a measure of primary production), B/P ratios (turnover time), and photosynthesis/respiration ratios. There are longstanding and reasonably straightforward methods for these metrics (Vollenweider, 1969); however, they have never become routine monitoring tools. Because these metrics are referred to explicitly in Odum's 1985 model of ecosystem response to stress, it may be very useful to include them in tests of the conceptual model presented in this report. This is because measures of productivity and respiration include rate, rather than simply a measure of biomass present in a given area at a given time.

Metrics for reproduction that apply across laboratory, mesocosm, and field studies would be very useful; however, the development of such metrics has lagged behind the development of the more familiar metrics such as mortality and growth. This is unfortunate, because reproduction is fundamental to the persistence of populations and is often the most sensitive indicator of stressor effects. Laboratory studies using zooplankton species measure reproduction as the number of young per female (e.g., the *Daphnia magna* and *Ceriodaphnia dubia* standard test methods). However, reproductive endpoints for other invertebrate test species that are more relevant to lotic systems (e.g., mayflies or chironomids) are not included in standard laboratory test methods. This is understandable, because the measurement of reproduction in insect species would require test durations that include the adult stage and test conditions that would accommodate adult emergence and egg-laying. The inclusion of reproductive endpoints for lotic species in both laboratory and mesocosm tests may be feasible, but would require some method development. The measurement of reproduction in field studies could include some well-established indirect measures such as emergence rate, using emergence traps (Edmondson and Weinberg, 1971). However, emergence is not a routine field measure and incorporation of this measure would have to be further developed and tested for reliability and effectiveness.

Structural endpoints that are applicable at all three scales of study are more challenging because of the single-species nature and small-scale of laboratory tests. However, the mortality endpoint common to most laboratory toxicity tests can be used in support of inferences about change in community structure, because highly sensitive species identified in the laboratory can be expected to decline in abundance more rapidly than less-sensitive species in mesocosm and field studies (Table 4-2). We propose a suite of metrics common to mesocosm and field studies. The use of a common set of metrics will facilitate comparisons among data obtained in the same geographical region.

Table 4-2. Sample of Effect Endpoints for Algae and Invertebrates Used in Laboratory, Mesocosm, and Field Studies That Support the Weight-of-Evidence Approach for Assessing Environmental Impacts of Multiple Stressors.

	Laboratory Experiments	Mesocosm Experiments	Field Observations
Exposures	Standard range-finding for individual stressors and stressor combinations	Dose-response curves for stressors and combination of stressors to determine appropriate range for dosing regimes Test combination of stressors on communities using laboratory	Field data along river gradient of stressor concentrations to compare with mesocosm results.
Test Organisms	Algal growth and survival* Invertebrate growth and survival* Single-species studies	Periphyton community. Invertebrate community. Experimental communities.	Periphyton community. Invertebrate community. *In situ* communities.
Functional Endpoints	Production or growth: biomass (cell counts), chlorophyll *a*, CO^2/O^2 production, growth rate Reproduction—eggs or young/female **	Production or growth: chlorophyll *a*, biomass, CO_2/O_2 production, organism size/weight, insect emergence **	Production or growth: biomass, chlorophyll *a*, insect emergence **
Structural Endpoints	Mortality—produces evidence in support of inferences about community structure.	Abundance, species richness, species evenness, EPT/Diptera, similarity indices.	Abundance, species richness, species evenness, EPT/Diptera, similarity indices.

* If possible, test species should be relevant to lotic systems (e.g., a diatom species rather than the standard green algae species, *Selenastrum capricornutum* and a mayfly species in addition to the standard water scud species, *Hyalella azteca*).

** Reproductive endpoints are not well-developed except for zooplankton and some marine invertebrate test species. Insect emergence is an indirect measure of reproductive capability and secondary production.

The suite of metrics for structural responses should include abundance (cell counts or individuals/m^2), species richness, similarity indices, species evenness, and Ephemeroptera/ Plecoptera/Trichoptera (EPT) /Diptera ratios.

Many of these metrics were developed for field studies (e.g., species richness, % EPT) and are readily transferable to the larger mesocosm designs that can accommodate periphyton or invertebrate communities. However, to reliably discover subtle disturbance effects, individual taxa must be examined with respect to their ecology and sensitivity.

The use of single summary indices that lump together meaningful metrics should be avoided when testing this multiple-stressor conceptual model. Combining various metrics into a single summary index creates arbitrary, ambiguous combinations of disparate measurements without any rationale of how and why they are combined and no clear idea of what the resulting index means. Such indices have no scientific justification beyond convenience, no objective units or scale, and no interpretive value, and, in fact, disguise or obscure the real changes occurring in the affected community. The metrics composing the index contain the useful information about real properties of benthic communities and putting them together into a single number is neither necessary nor scientifically justified. Ecosystems are complex and any attempt to compress the response of the whole community into one number is necessarily going to discard a great deal of information.

There are several effects endpoints for fish species that apply across all three study types; however, the incorporation of fish into mesocosm studies introduces a significant additional level of logistical complexity and cost. Therefore, the study team recommends that the initial series of tests of the conceptual model focus on algae and invertebrate test species. In the future, whole-organism effects endpoints for fish that will be useful will include:

- Mortality—applicable to laboratory and mesocosm
- Reproductive endpoints (fecundity, relative gonad size)—applicable to laboratory, mesocosm, and field studies
- Growth (age/size relationships)—applicable to laboratory, mesocosm, and field studies
- Health and condition indices (condition factor; gross pathology, histopathology, relative liver size)

These endpoints can be supplemented by population-level endpoints collected from the field such as relative abundance, age-to-maturity, age and size distribution, and recruitment of young-of-the-year. A considerable body of experience using these fish effects endpoints has been developed under the Canadian Environmental Effects Monitoring (EEM) regulations (Environment Canada, 1998) for the pulp and paper industry (and now applicable to metal mining). The EEM data provide valuable information on the variability of the endpoints and the statistical power required in a study design to detect differences from reference populations in field studies. The EEM database also provides complementary laboratory test data (acute and chronic) that can be compared to field results. Information on the EEM program and access to data can be obtained by contacting Environment Canada.

Scaling factors are incorporated into the study design proposed for testing the conceptual model by the use of a combination of laboratory, mesocosm, and field approaches. This combination of approaches allows the “scale-up” from small-to-medium scale via the comparison of laboratory and mesocosm results and from medium-to-large scale via the comparison of mesocosm, and field results. This scaling can involve both temporal (short-term versus long-term) and spatial scales.

Lowell and Culp (2002) have shown that adequate sampling programs that ensure sufficient replication, have the statistical power to detect spatial effects associated with exposure to point-source complex discharges and adequate sampling frequency will ensure that temporal effects are accounted for.

These considerations apply equally to both medium- and large-scale mesocosms, as carefully designed medium-scale mesocosm systems are complex, multispecies systems that can provide a similar level of “realism” to larger-scale mesocosm systems.

4.4 Assembling the Weight of Evidence

The application of the General Test Scheme to the testing of the three risk hypotheses for multiple stressors will generate several lines of evidence from laboratory, mesocosm, and field studies. The process of determining whether the risk hypotheses should be accepted or rejected will require a formal assessment of all lines of evidence.

4.4.1 Defining Weight of Evidence

In risk assessment terminology, weight of evidence is "the process by which multiple measurement endpoints are related to an assessment endpoint to evaluate whether significant risk is posed to the environment" (Massachusetts Weight-of-Evidence Work Group, 1995). This definition can be amended to state that "weight of evidence is the process by which laboratory, mesocosm, and field effects endpoints are evaluated with respect to their concordance with the multiple-stressor risk hypotheses."

Three characteristics of endpoints are critical to the weight-of-evidence process in risk assessment: the weight assigned to each measure; the magnitude of response observed in the measure; and the concurrence among multiple measures (Massachusetts Weight-of-Evidence Work Group, 1995). Consideration must also be given to the adequacy and quality of the data, the degree and type of uncertainty associated with each line of evidence, and the relationship between the evidence and the risk assessment objectives (U.S. EPA, 1998).

The above critical characteristics of endpoints are not necessarily applicable to the testing of the multiple-stressor risk hypotheses. The weight assigned to each measure in risk assessment often involves considerable professional judgment and usually incorporates criteria such as the strength of the evidence for cause/effect relationships and ecological relevance. The study team suggests that there be no weight applied to the lines of evidence produced by the General Test Scheme, because each line of evidence will have a balance of strengths and weaknesses. The magnitude of response and the concurrence among multiple measures are important criteria; however, there are several additional criteria that will assist in the evaluation of the weight of evidence. These criteria are listed and described below. Of course, the adequacy and quality of the data will always be a required criterion. The degree and type of uncertainty associated with each line of evidence should be described in terms of the three primary sources of uncertainty: 1) measurement error (precision); 2) natural variability (accuracy); and 3) model error (the degree to which the line of evidence may or may not conform to natural processes).

The study team has determined that there are three critical steps in assembling the weight of evidence from the General Test Scheme for testing the multiple-stressor risk hypotheses: 1) establishing causality, 2) defining relative risk, and 3) assembling the lines of evidence in a decision-making framework. These three steps are derived from Lowell et al., (2000), and have been used successfully in testing and refining conceptual models of the effects of multiple stressors in large western Canadian rivers.

Step One: Establishing Causality

The triad of field, laboratory and mesocosm data produced by the General Test Scheme must be examined for cause/effect relationships between the stressors and endpoints. Without strong evidence for causality, the multiple-stressor hypotheses cannot be successfully tested. Distinguishing among the cumulative effects of multiple stressors (which sometimes have confounding effects) requires establishing a strong causal link to each stressor. This requires evaluation using a formal set of criteria. Several authors have suggested causality criteria, all based on the work of epidemiologists and ecologists over the past 30 years (e.g., Hill 1965, Fox, 1991). The criteria presented in Table 4-3 are based on a combination of those assembled by Lowell et al. (2000) and those presented by the U.S. EPA (2000) in their SI Guidance Document. The types of effects endpoints and whether they are generated by laboratory, mesocosm or field studies are matched with the causality criteria in Table 4-3.

Each causality criterion is briefly described below.

Spatial Correlation: The spatial co-location of the candidate cause and the effect endpoint in field data provides evidence for a cause/effect relationship. However, causes occurring upstream may mask the effects of causes occurring later in the downstream sequence.

Temporal Correlation: A cause must always precede its effects when examining field data. Causes occurring early in a time sequence may mask the effects of causes occurring later.

Biological Gradient/Strength: Effects should increase with increasing exposure. This criterion applies to laboratory, mesocosm, and field lines of evidence. This is a classic toxicological requirement and does not acknowledge the possibility that multiple stressors may have subsidy effects followed by toxicant effects and that effects may be additive, synergistic, or antagonistic. Therefore, this criterion should be applied with caution in multiple-stressor situations and only in the context of the multiple-stressor risk hypotheses. The strength of the response may be the more useful aspect of this criterion. Regressions and correlation analyses can help define the strength of the response by providing regression and correlation coefficients. Separate analyses may be required for different regions of the subsidy-toxicant response curve.

Evidence of a Complete Exposure Pathway: If the stressor does not reach the receptor, it cannot cause an effect. Exposure must be demonstrated if the risk hypotheses are to be tested. Measures of exposure must be part of laboratory, mesocosm, and field studies.

Table 4-3. Types of Evidence (Columns) that Contribute to Each Causal Criterion (Rows) Adapted from U.S. EPA (2000) and Lowell et al (2000).

Causal Criterion	Types of Evidence												
	Associations of Measurements of Cause and Effect				Measurements Related to Causal Mechanisms					Case Exposure/ Other Exposure Response		Experiments	
	Spatial Co-location Field)	Spatial Gradient (Field)	Temporal Co-occurrence (Field)	Temporal Gradient (Field)	Symptoms & Biomarkers (Field Mesocosms, Lab)	Structural & Functional Endpoints (Field, Mesocosms, Lab)	Indirect Effects (Field, Mesocosm)	Effects on Other Receptors (Field)	Cause and Receptor are Co-located (Field)	Laboratory Exposure Response (Lab and Mesocosm)	Field-derived Exposure Response (Field)	Manipulation of Sources (Field, Mesocosm, Lab)	Manipulation of Exposure (Lab, Mesocosm)
Spatial Correlation	X	X							X			X	
Temporal Correlation			X	X								X	
Biological Gradient/ Strength		X		X	X	X				X	X	X	X
Consistency of Association	X		X					X		X		X	X
Evidence of Complete Exposure Pathway					X	X	X	X	X				
Specificity of Cause					X	X				X		X	X
Plausibility: Mechanism					X	X	X	X	X				
Plausibility: Stressor-Response					X	X				X	X	X	X
Experimental Verification										X		X	X
Coherence with Analogous Cases	X	X	X	X						X	X		

Consistency of Association: The repeated observation of the effect and the candidate cause (or the experimentally applied stressor) in different field studies or in different experiments builds a case for causation. If laboratory, mesocosm, and field effects endpoints show a consistent pattern of response to the multiple-stressor combination being studied, and that pattern is consistent with the predictions generated by the conceptual model and risk hypotheses, then the case for accepting the risk hypotheses is strengthened.

Specificity of Cause: If an effect has only one or a few known causes, then the occurrence of one of those causes in association with the effect is strong evidence of causation. For example, blue-sac disease in fish is associated with particular organic compounds such as PCBs, some polycyclic aromatic hydrocarbons (PAHs), and dioxins and furans. Unfortunately, strict specificity is rare.

Plausibility for the Mechanism: There should be a plausible mechanism known from past experience or from general biological knowledge for the cause to produce the observed effects. For example, there are well-established mechanisms for the direct and indirect effects of nutrients. However, there may not be well-established mechanisms for the confounding effects of nutrients on metal toxicity (although plausible theories such as the Biotic Ligand Model may apply).

Plausibility for the Stressor-Response: The level of stressor applied in laboratory and mesocosm experiments or observed in the field should be in the range where effects would be expected, given our knowledge of the stressor-response relationship. This criterion must be applied in the context of the predicted subsidy-toxicant response curve to test the risk hypotheses.

Coherence with Analogous Cases: The results of laboratory, mesocosm, and field studies for a particular combination of stressors should be examined for similarity to other cases. This may be the most difficult criterion to apply to the multiple-stressor risk hypotheses, because it is likely that no two multiple-stressor situations will be sufficiently similar. However, examination for consistency with the predicted subsidy-toxicant response curve across several reasonably similar multiple-stressor situations would help build confidence in the multiple-stressor conceptual model.

The results of the examination for causality among the laboratory, mesocosm, and field lines of evidence can be summarized by applying scores that reflect the "performance" of each line of evidence against the causality criteria. An example of a scoring scheme is presented in Table 4-4, and is adapted from U.S. EPA (2000). Scoring schemes assist in the presentation of the results and can be carried forward to the final weight-of-evidence framework.

Step Two: Defining Relative Risk

The weighing of evidence involves summarizing the available lines of evidence for each endpoint against some benchmark of relative risk. The definition of relative risk should be as clear and explicit as possible. The definition will always involve the use of professional judgment; however, evidence in support of the definition from the scientific literature should be provided.

Table 4-4. Example of a Scoring Scheme for Causal Criteria (Adapted from U.S. EPA, 2000).

Criterion	Results	Score[1]
Spatial Correlation	Compatible, Uncertain, Incompatible	+, 0, – – –
Temporal Correlation	Compatible, Uncertain, Incompatible	+, 0, – – –
Consistency of Association	Invariant, In many places and times, At background frequencies or many exception to the association	++, +, –
Biological Gradient/Strength	Strong and monotonic, Weak or other than monotonic, None, Clear association but wrong sign	+++, +, –, – – –
Evidence of Complete Exposure Pathway	Evidence for all steps, Incomplete evidence, Ambiguous, Some steps missing or implausible	++, +, 0, –
Experimental Verification	Experimental studies: Concordant, Ambiguous, Inconcordant	+++, 0, – – –
Plausible Mechanism	Actual Evidence, Plausible, Not known, Implausible	++, +, 0, –
Plausible Stressor-Response[2]	Quantitatively consistent, Concordant, Ambiguous, Inconcordant	+++, +, 0, –
Specificity of Cause[3]	Only possible cause, One of a few, One of many	+++, ++, 0
Analogy Coherence:	Analogous cases:	
Positive	Many or few but clear,	++, +
Negative	Few or unclear	– –, –

[1] In addition to the scores noted, there may be no evidence (NE) available relevant to the consideration, or the consideration may be not applicable (NA) for the particular case (see especially stressor-response and specificity).

[2] Stressor-response is not applicable (NA) if the mechanism is clearly implausible.

[3] Specificity of cause is not applicable (NA) if either the mechanism is clearly implausible, or if there are many exceptions to the association.

The most commonly cited ecological benchmark for the definition of *de minimis* risk is "less than 20% reduction in the abundance or production of an endpoint population" (Suter et al., 1995). Suter et al. (1995) continue with two additional risk benchmarks: 1) loss of less than 20% of the species in an endpoint community in a unit area, and 2) loss of less than 20% of the area of an endpoint community in a unit area. Unit area refers to a discrete area that is at risk and may be subject to a regulatory or remedial action. Loss of more than 20% of a community may be considered acceptable if the community has negligible ecological value (e.g., a baseball field) or if the loss is brief because the community is adapted to physical disturbance (e.g., the plant communities of stream gravel bars) (Suter et al., 1995).

Suter et al. (1995) confronted the issue of defining "acceptable" risk by reviewing the then-current regulatory precedents in the United States. The 20% effect rule has its history with United States regulatory policy (e.g., development of National Ambient Water Quality Criteria, the Office of Pesticide Programs pesticide testing with birds), but also recognizes the power of aquatic toxicity tests (i.e., tests used to regulate effluents cannot reliably detect less than a 20% reduction in the test endpoints). The *de minimus* policy uses 20% as the approximate limit of detection of aquatic field measurement techniques (e.g., benthic invertebrate community metrics) (Swanson and Hull, 2003).

This 20% level was the first to be defined, and has since been used in ecological risk assessments and other programs in the United States and Canada. A version of the same reasoning is being used in Canadian EEM programs. In EEM, a "critical effect size" is established. Methods for determination of critical effect sizes are still a subject of considerable scientific discussion; however, a common starting point is the normal range of variability seen in reference communities (Washington State Department of Ecology, 1991; Elliot, 1996; Lowell, 1997). This is often defined as two standard deviations from the reference mean. However, the 20% benchmark assumes some ability to distinguish 'at risk' communities according to measured differences with some sort of reference condition. The validity of this comparison may be disputed (Swanson et al., 2004).

For a particular multiple-stressor study implemented using the study design in this report, the 20% level can be a starting point for definition of *de minimis* risk; however, the benchmark will vary with receptor, endpoint, and regulatory context. For example, a 20% change in some fish health measures (e.g., gonad size) can be detected using achievable and practical sample sizes, and has been shown to be meaningful with respect to population-level effects. However, a 20% change in the abundance of particular invertebrate species is often well within the range of natural variability. Therefore, the "critical effect size" for the various laboratory, mesocosm, and field endpoints must be determined using available literature, existing baseline data, and evaluation of the relevance to the multiple-stressor conceptual model.

The example of a set of relative risk rankings presented by Chapman (2003) in Table 4-5 illustrates how a diverse set of endpoints can be evaluated. The definitions of "high," "moderate," and "low" risk vary considerably with each type of endpoint. This example is for a situation involving sediment contamination with mercury and organics. The weight of evidence was used as the basis for sediment remediation recommendations. The benchmarks for each endpoint were the result of debate and discussion among scientists and stakeholders and should not be extrapolated to other situations without careful consideration of their applicability.

Benchmarks for WERF-sponsored testing of the multiple-stressor conceptual model must be useful for evaluating the three multiple-stressor risk hypotheses. Therefore, it must be possible to translate the relative risk rankings back to positions on the hypothesized subsidy-toxicant response curve and to the predicted endpoint responses in the linkage diagrams. An example of an applicable risk ranking for testing the conceptual model using algal effects endpoints is presented in Section 4.4.3.

Step Three: Summing Up the Weight of Evidence in a Decision-Making Framework

The process of weighing the evidence amounts to determining what overall risk statement is best supported by the individual lines of evidence (Suter, 1996). If the benchmarks for relative risk have been carefully and clearly defined, the production of a summary weight-of-evidence table should be quite straightforward. The "art" involved in summary tables is the development

Table 4-5. Example of a Set of Relative Risk Rankings (extracted from Chapman, 2003).

Example 1 (Great Lakes)	O	o	*
Chemistry (Bulk sediment)	Exceeds upper-bound SQG[1] (>SEL[2])	Exceeds lower-bound SQG (>LEL[3])	All contaminant concentrations below lower-bound SQG (>LEL)
Toxicity	EC50 (Study 1); 50% of "warning level" from Great Lakes references (Study 2)	EC20 (Study 1); Less than the "warning level" (Study 2)	Negligible effects compared to references, controls, and/or "warning level"
Bioaccumulation	10-fold difference compared to references	2-fold difference compared to references	No increase in contaminant uptake
Benthos Assemblages (multivariate assessment)	"different" or "very different" from reference stations	"possibly different" from reference stations	"equivalent" to reference stations
Overall Risk	***High***—Significant adverse effects predicted	***Moderate***—Effects possible, but unlikely and/or small in magnitude	***Low***—No significant adverse effects predicted

[1] SQG = Sediment Quality Guideline
[2] SEL = Severe Effects Level
[3] LEL = Lowest Effects Level

of logical statements in support of the "overall risk ranking" derived from the summation of relative risk scores for all endpoints.

The example of a weight-of-evidence summary table shown in Table 4-6 shows how the overall risk ranking can appear to be reasonably obvious and defensible, even when a rather complex set of endpoints and risk ranks are involved. This example was used to communicate the results of a weight-of-evidence risk characterization of contaminated sediments to stakeholders. The stakeholders were generally satisfied with the method and used the weight-of-evidence results in their further deliberations regarding sediment remediation options. The results in Table 4-6 must be accompanied by the definition of risk ranks in Table 4-5, or the logic supporting the overall risk ranking will not be obvious.

Suter (1996) suggests that the logical analysis of the lines of evidence should proceed from the most realistic (i.e., field studies) to the most precise and controlled (e.g., laboratory toxicity tests). That is, field surveys indicate the actual state of the environment, and although there are limitations to field data, provide a strong line of evidence. Toxicity tests provide an indication of whether toxicity may be responsible for effects in the field, but also have limitations, including their short duration. Single-chemical toxicity data (e.g., data used in risk modeling) generally provide the weakest line of evidence, because they are not based on site-specific conditions. However, all of these lines of evidence combined can be used to evaluate the significance of risk to an assessment endpoint. The application of Suter's suggested method is illustrated in Table 4-7. However, as pointed out earlier, the study team does not recommend the application of weights to the various lines of evidence arising from the General Test Scheme for multiple stressors. This is because the combination of laboratory, mesocosm, and field approaches will be planned carefully, in advance, so to take advantage of the strengths of each type of study.

The use of symbols, numeric scores, or narratives within weight-of-evidence summary tables is at the discretion of the study team. Symbols can be very effective in risk communication; however, as noted above, the logic in support of each of the symbols must be clearly explained. The same provision applies to the use of numeric scores, particularly if some lines of evidence are weighted more heavily than others. Narratives reduce the necessity for a series of supporting tables; however, they can produce lengthy and complex summary tables that are very difficult to follow.

4.4.2 Using the Weight-of-Evidence Approachó Examples from the Literature

There are very few examples in the literature that have used a weight-of-evidence approach to assess the effects of multiple stressors in real life examples. Typically, these either do not address all critical steps used for assembling weight of evidence or do not use very clearly defined steps in doing so. Thus, there is a pressing need for the development of a clear, direct and concise method that can be used by researchers and regulators to best "quantify risk" associated with multiple stressors.

Here we provide a brief overview of two sets of work that have focused on evaluating several lines of evidence in multiple-stressor situations.

- The recent work in large Canadian river ecosystems integrating mesocosms, laboratory, and field studies (Lowell et al., 2000; Culp et al., 2000a).
- The SI Guidance Document by the U.S. EPA (2000).

Although we acknowledge their considerable contribution, we point out areas where further work or clarity is still required.

The work of Culp et al. illustrates the use of integrated laboratory, mesocosm, and field studies for a weight-of-evidence approach to the assessment of multiple-stressor effects on river ecosystems. These researchers have focused on using mesocosm systems to bridge between the more commonly used laboratory testing and field biomonitoring. Culp et al. provide a detailed example of applying causal criteria to data from laboratory, mesocosm, and field studies of large rivers (Table 4-8).

The SI guidance document produced by the U.S. EPA is a comprehensive document that aims at determining the causes (SI) for the observed effects in ecosystems containing multiple stressors. It relies heavily on availability of good-quality field data complemented by laboratory experiments. The guidelines contained in the document are very valuable for such cases where there is a need to identify the stressor (or combination of stressors) for the observed effects and extensive field data are available.

Table 4-6. Summary of Relative and Overall Risk Ranking Using a Weight-of-evidence Approach.

Site	Bulk Sediment Chemistry: Metal (1,2) - PAH - PCB	*Chironomus* Toxicity: Survival (1,2) - Growth (1,2)	*Hexagenia* Toxicity: Survival (1,2) - Growth (1,2)	*Hyalella* Toxicity Survival Growth	Fathead Minnow: Survival - Bioaccum.	*Tubifex* Toxicity: Survival - Cocoons % Hatch No. Young	Multivariate Evaluation of Benthic Community Structure	Overall Risk Rating
1	o-o-o-o	*-*-*-*	*-*-*-*	*-*	na-na	*-*-o-o	*	Low
2	O-O-o-o	*-*-*-*	*-*-*-*	*-*	na -na	*-*-*-*	*	Low
3	na-o-na-na	na-*-na-*	na-*-na-*	na-na	*-*	*-*-*-O	*	Low
4	na-O-na-na	na-*-na-*	na-*-na-*	na-na	*-*	*-*-o-O	o	Moderate
5	O-O-O-o	*-*-*-*	*-*-*-*	*-*	na-na	*-*-*-o	*	Low
6	O-O-o-o	*-*-o-*	*-*-o-*	*-*	na-na	*-*-*-*	*	Low
7	na-O-na-na	na-*-na-*	na-*-na-*	na-na	*-*	*-*-*-*	*	Low
8	o-o-o-o	o-*-*-*	*-*-*-*	*-*	na-na	*-*-o-O	*	Low
9	na-O-na-na	na-*-na-*	na-*-na-*	na-na	*-*	*-*-o-o	o	Low
10	o-o-o-o	*-*-*-*	*-*-*-*	*-*	na-na	*-*-o-O	*	Low
Ref A	o-o-*-*	na-*-na-*	na-*-na-*	*-*	na-na	*-*-*-*	o	Low
Ref B	na-o-na-na	na-*-na-*	na-*-na-*	na-na	*-*	*-*-*-*	o	Low

Table 4-7. Sample Risk Characterization Summary Using Pre-Assigned Weights (from Suter, 1996).

Evidence	Result[1]	Weight[2]	Explanation
Biological Surveys	-	H	Fish community productivity and species richness are both high in reach 2, relative to upstream reaches. Data are abundant and of high quality.
Ambient Toxicity Tests	$\pm$	M	High lethality to fathead minnow larvae in a single test at Site 3.3, but variability is too high for standard statistical significance. No other aqueous toxicity was observed.
Water Analysis/Single Chemical Tests	$\pm$	M	Only Zn is believed to be potentially toxic in water and only to highly sensitive species. Limited data.
Weight of Evidence	-		**Reach 2 supports a clearly high-quality fish community. Other evidence that suggests toxic risks is much weaker (single chemical toxicology) or inconsistent and weak (ambient toxicity test)**

[1] + indicates that the evidence is consistent with the occurrence of the endpoint effect
- indicates that the evidence is inconsistent with the occurrence of the endpoint effect
$\pm$ indicates that the evidence is too ambiguous to interpret

[2] H indicates a high weighting, or greater confidence in the line of evidence
M indicates a moderate weighting, or an intermediate level of confidence in the line of evidence
L indicates a low weighting, or lower confidence in the line of evidence

The two sets of work make use of the three main steps involved in assembling the weight of evidence in a risk assessment framework. The three main steps are:

1. Establishing causality
2. Defining relative risk
3. Summing up the weight of evidence in a decision-making framework.

A summary of how Culp et al. and the SI guidance approach the problem of assembling the weight of evidence for multiple stressors is provided in Table 4-9.

These two examples illustrate the challenges inherent in assembling a weight of evidence in multiple-stressor situations. A complete review of examples of weight-of-evidence approaches is not within the scope of this report. To demonstrate how the three-step weight-of-evidence process would apply to testing of the conceptual model for multiple stressors presented in this report, a hypothetical example is presented below.

4.4.2.1 Example of Applying the Three-Step Process to Evidence from Multiple-stressor studies

The three-step process for assembling a weight of evidence can be illustrated using a hypothetical set of results of laboratory, mesocosm, and field investigations of copper plus phosphorus effects on algae. The hypothetical results are the same as those presented in the Linkage Diagrams for copper plus phosphorus examples in Chapter 3.0. The method illustrated by this example is not a rigid guide; instead, it is one suggested approach for assembling, interpreting and presenting the evidence in a transparent manner.

Table 4-8. Causal Criteria and Support for a Weight-of-Evidence Approach.

Causal Criterion	Support for Criterion
Spatial correlation of stressor and effect along gradient from more to less exposed areas.	Gradient in effluent/phosphorus concentration down river from pulp mill/sewage treatment plant positively correlated with algal biomass in river.
Temporal correlation of stressor and effect relative to time course of exposure.	Benthic invertebrate community structure correlated with pulp mill output of effluent solids and phosphorus over 20-year period. Algal biomass and chironomid abundance traced recent upsets/changes in pulp mill effluent treatment.
Plausible mechanism linking stressor and effect	Increases in growth and abundance, measured in several studies, consistently agreed with known mechanisms for phosphorus/organic matter nutrient enhancement. Negative low DO effects also consistent with expected physiological responses.
Experimental verification of stressor effects under controlled conditions and concordance of experimental results with field data.	In-river nutrient additions and riverside artificial stream experiments verified (and clarified mechanisms for) nutrient enhancement hypotheses, in agreement with field data. Negative low-DO effects also experimentally verified.
Strength: steep exposure - response curve.	Artificial stream experiments demonstrated pronounced increases in algal/invertebrate growth with small increases in effluent concentration.
Specificity: effect diagnostic of exposure to a particular stressor.	Increases in growth, measured in several studies, characteristic of nutrient enhancement effects of phosphorus/organic matter additions to oligotrophic rivers.
Evidence of exposure (contaminants or other indicators) in body of affected organisms.	Stable isotopic analyses indicated presence of terrestrially derived (probably pulp mill effluent) carbon and nitrogen in invertebrates and fish down-river from mill.
Consistency of stressor/effect association among different studies within the region being studied.	Nutrient enhancement effects of effluent apparent in all studies conducted, including temporal and spatial field correlations and results from field and streamside experiments. Some toxicity apparent at higher concentrations in two sets of experiments and in field following upsets in effluent treatment.
Coherence with existing knowledge from other regions where the same or analogous stressors and effects have been studied.	Similar nutrient enhancement effects observed in both Thompson and Athabasca River basins. Nutrient enhancement and low DO effects agree with the results of studies conducted in other regions.

Reprinted with permission from Lowell et al., 2000. Causal criteria are derived from Fox, 1991, Suter, 1993, Beyers, 1998, Lowell et al., 2000, Culp et al., 2000a). Copyright SETAC, Pensacola, FL, USA.

Step One: Examine Causal Criteria

The outcome of this step would be a ranking of the strength of the evidence for causation for each line of evidence (Table 4-10). To provide complete support and transparency, Table 4-10 would be supported by Appendix tables presenting the performance of each line of evidence against each causal criterion. These appendix tables are not provided in this example.

Step Two: Define Relative Risk

The hypothetical example for defining risk presents benchmarks for interpreting the nature and strength of the response to the two stressors, copper and phosphorus. Because phosphorus can produce a subsidy response whereas copper primarily produces a toxicant response, the definition of the nature of the response must include both stimulation and inhibition.

Table 4-9. Summary of Two Different Approaches for Assembling Weight of Evidence.

	Assembling the Weight of Evidence					
	Step 1 (Establishing Causality)		Step 2 (Determining Relative Risk)		Step 3 (Assembling the Weight of Evidence)	
	Culp and Associates[1]	**SI**[2]	**Culp and Associates**	**SI**	**Culp and Associates**[3]	**SI**[4]
trengths	Comprehen-sive, concise —uses one summary table with a clear narrative addressing each causality criterion	comprehensive, provides ranking to causality criteria, produces a through documentation of evidence for causality.	This step is incorporated with step 3	N/A (The SI process is designed to show causation, not to rank risks).	Extensive, integrates data from laboratory, mesocosm, and field studies to produce refined conceptual model of multiple-stressor effects.	Extensive, detailed analyses of the various lines of evidence originating from field and lab studies.
eaknesses	No ranking of causality criteria; therefore, the reader must rely on the narratives.	Lengthy process, numerous tables with accompany-ing text, can be very difficult to follow. Works well only when extensive database available.	No defined benchmarks for "relative risk"	N/A	Narrative, no symbols used, no defined benchmarks for relative risk. Does not produce an overall risk ranking.	Strictly for causal characterization—Lengthy process with complex results and does not produce an overall risk ranking

[1] Table 1 in (Lowell et al., 2000).

[2] Table 7.9 and accompanying tables in section 4.2.3 (U.S. EPA, 2000).

[3] Table 2 in (Culp et al., 2000a).

[4] Table 7.12 in section 7.10 (U.S. EPA, 2000).

Table 4-10. A Hypothetical Example of Ranking Lines of Evidence Against Causal Criteria: Copper Plus Phosphorus Effects on Algal Endpoints.

Line of Evidence	Causal Rank	Rationale
Laboratory		
Growth	+++	Strong, consistent, plausible, stressor concentration/response relationship showing expected stimulation with phosphorus alone, inhibition with copper alone, and subsidy effects being overcome by inhibition at higher copper concentrations
Mortality	+++	Strong, consistent, plausible, stressor concentration/response relationship showing expected lack of mortality with phosphorus alone, mortality of copper-sensitive test species with copper alone, and subsidy effects being overcome by mortality at higher copper concentrations
Mesocosm		
Chlorophyll *a*	++	Relatively strong, consistent, plausible stressor concentration/response relationship showing expected increase in chlorophyll *a* with phosphorus alone, decrease with copper alone, and subsidy effects being overcome at higher copper concentrations
Dry weight	++	Relatively strong, consistent, plausible stressor concentration/response relationship showing expected increase in dry weight with phosphorus alone, decrease with copper alone, and subsidy effects being overcome at higher copper concentrations
Density (cells/m2)	++	Relatively strong, consistent, plausible stressor concentration/response relationship showing expected increase in density with phosphorus alone, decrease with copper alone, and subsidy effects being overcome at higher copper concentrations causing a decrease in density, especially of sensitive species
Field		
Chlorophyll *a*	+	Regression and correlation coefficients of 0.5 or less between phosphorus and copper concentrations and chlorophyll *a*, illustrating some evidence for spatial and temporal correlation but with high variability; multivariate analysis shows strong influence of confounding variables such as light and flow, lack of strong indication of switch from subsidy to toxicant effect
Dry weight	+	Regression and correlation coefficients of 0.5 or less between phosphorus and copper concentrations and dry weight, illustrating some evidence for spatial and temporal correlation but with high variability; multivariate analysis shows strong influence of confounding variables such as light and flow, lack of strong indication of switch from subsidy to toxicant effect.
Density (cells/m^2)	+	Regression and correlation coefficients of 0.3 or less between phosphorus and copper concentrations and density, illustrating weak evidence for spatial and temporal correlation with evidence of increased density of copper-tolerant species, multivariate analysis shows that most of the variance is accounted for by variables other than phosphorus and copper

The multiple-stressor conceptual model shows four main “regions” in the subsidy-toxicant response curve (Figure 3-3):

[a] region: subsidy effects with functional responses occurring first
[b] region: the “cusp” between subsidy and toxicant effects with variable functional and structural responses
[c] region: mild-moderate toxicant effects with structural responses occurring first and mild to no functional responses because of functional redundancy

[d] region: strong toxicant effects with structural responses occurring first followed by moderate to strong functional response because of the loss of functional redundancy.

The definition of relative risk is based on the above four "regions" of the subsidy-toxicant response curve, plus consideration of the strength of the causal relationships between stressor concentrations and measurement endpoints.

Relative risk is defined using symbols to represent the various combinations of response-plus-causal rank in Table 4-11. The circle symbols represent toxicant responses in the [c] and [d] regions of the subsidy-toxicant response curve. The square symbols represent the "cusp" [b] region of the curve. The triangle symbols represent the subsidy [a] region of the curve. The degree of shading corresponds to the strength of the causal relationship, from "full" symbols representing a strong causal relationship to "void" symbols representing a weak one.

Table 4-11. Hypothetical Ranking System for Relative Risk Plus Causality.

Relative Risk	High Causal Rank	Moderate Causal Rank	Low Causal Rank
Strong toxicant: [d] range of curve	●	◕	○
Moderate toxicant: [c] range of curve	◕	◐	○
Cusp between toxicant and subsidy: [b] range of curve	■	◧	□
Subsidy: [a] range of curve	▲	◭	△

● = high causal rank, strong toxicant in [d] region, ◕ = moderate causal rank, strong toxicant in [d] region or high causal rank, moderate toxicant in [c] region, ◐ = moderate causal rank, moderate toxicant in [c] region, ○ = low causal rank, strong toxicant in [d] region

■ = high causal rank, toxicant/subsidy cusp in [b] region, ◧ = moderate causal rank, toxicant/subsidy cusp in [b] region, □ = low causal rank, toxicant/subsidy cusp in [b] region

▲ = high causal rank, subsidy in [a] region, ◭ = moderate causal rank, subsidy in [a] region. △ = low causal rank, subsidy in [a] region

Alternatives to the use of symbols would include numeric scores or narratives. A potential drawback of numeric scores is that they imply a greater level of precision and accuracy than is actually the case. Narratives allow for a more thorough description of all lines of evidence, but can be confusing when several permutations and combinations of relative risks x causal ranks are involved.

Once a general system for describing relative risk is developed, a system for matching quantitative results for each endpoint with the ranking system must be derived. Professional judgment is required to place quantitative results within qualitative rankings such as "weak," "moderate," or "strong." The rationale behind the judgment must be clearly described in the accompanying text.

The responses of laboratory, mesocosm, and field studies of algal responses to copper and phosphorus could be ranked as illustrated in Table 4-12. The benchmarks for the various ranks are based on professional judgment regarding the type of data that would be associated with strong or moderately toxicant results, or with subsidy response. For example, if chlorophyll *a*, dry weight, and algal density all show an increase that falls outside the range of natural variability (which is assumed to be 20%), then changes that are greater than 120% of upstream reference sites would be interpreted as a subsidy response.

The type of risk ranking table represented by Table 4-12 is an excellent focal point for discussions with stakeholders and regulators. The suggested rationale underlying the risk rankings can be discussed to achieve a consensus regarding what constitutes a strong toxicant response (which would be associated with high risk) versus what constitutes a subsidy response (which may not be associated with any risk, depending upon baseline conditions). The benchmarks for definition of each risk rank will be case-specific, but could follow a common framework such as the one suggested in the hypothetical example.

Step Three: Assembling the Weight of Evidence

An overall weight-of-evidence table is produced by assigning a risk rank to each line of evidence, portraying the risk ranks, and then summing the risk ranks with a numeric score, symbol or a narrative. Table 4-13 shows how the hypothetical lines of evidence from studies of copper plus phosphorus would be assembled and interpreted. The treatments are the same as those used to illustrate the conceptual model in Chapter 3.0. Cross-references to the figures portraying the theoretical results from each treatment are provided in Table 4-13.

Overall effects are briefly described as a narrative in Table 4-13. The results in Table 4-13 correspond with what would be expected if the conceptual model holds true. Thus, for example, the combination of high copper and high phosphorus yields results that fall within the toxicant portion of the curve, whereas the combination of low copper and high phosphorus yields highly variable results that bracket the "cusp" of the subsidy-toxicant response curve. If results were not in agreement with the conceptual model, then a new iteration of the General Test Scheme (Figure 4-1) would take place (i.e., the model would be adjusted and another set of experiments and field observations would be conducted).

Table 4-12. Ranking Relative Risk for Multiple-stressor Studies Using Ecological Benchmarksó A Hypothetical Example with Phosphorus and Copper.

Line of Evidence	Relative Risk*			
	Strong Toxicant [d] Region of Curve	Moderate Toxicant [c] Region of Curve	Cusp Between Toxicant and Subsidy [b] Region of Curve	Subsidy [a] Region of Curve
Laboratory**				
Growth	EC50 # 25%	EC50 # 50%	EC50 # 100%	EC50 > 100%
Mortality	LC50 # 10%	LC50 # 25%	LC50 > 100%	LC50 > 100%
Mesocosm				
Chlorophyll *a*	<25% of Control	<50% of Control	>50%; <110% of Control	>100% of Control
Dry Weight	<25% of Control	<50% of Control	>50%; <110% of Control	>100% of Control
Density	<25% of Control	<50% of Control	>50%; <110% of Control	>100% of Control
Field				
Chlorophyll *a*	<25% of Upstream Reference	<50% of Upstream Reference	No Statistical Differences with Upstream Reference	>120% of Upstream Reference
Dry Weight	<25% of Upstream Reference	<50% of Upstream Reference		>120% of Upstream Reference
Density	<25% of Upstream Reference	<50% of Upstream Reference		>120% of Upstream Reference

* Degree of toxicant or subsidy effect defined by regions of the subsidy-toxicant curve (see Figure 3-5).

** Assume that laboratory results are expressed as a dilution of a starting test solution containing particular combinations of copper and phosphorus (e.g., high copper, low phosphorus).

Table 4-13. A Hypothetical Example of Assembling a Weight of Evidence from Multiple-stressor Studies Where the Results Correspond with What Would be Expected if the Conceptual Model Holds True.

Treatments	Laboratory		Mesocosm			Field			Overall Effect
	Growth	Mortality	Chlorophyll *a*	Dry Weight	Density	Chlorophyll *a*	Dry Weight	Density	
High Copper High Phosphorus (Figure 3-13)	●	●	●	◕	◕	◐	◐	○	Toxicant effect Strong structural response
High Copper Low Phosphorus (Figure 3-10)	●	●	◕	◕	◕	◐	◐	◐	Toxicant effect Some structural response Maintenance of function
Low Copper High Phosphorus (Figure 3-12)	◕	◕	■	◧	◧	◭	◭	△	Highly variable responses Primarily subsidy effects observed in the field
Low Copper Low Phosphorus (Figure 3-9)	■	■	◧	◧	◭	△	△	△	Few measurable effects Subsidy and toxicant effects balanced

The example presented above shows how data generated by the General Test Scheme for the conceptual model can be interpreted within a risk-based weight-of-evidence framework. Actual data will be more complex, and will involve a longer list of endpoints. However, if the basic three-step process is carefully followed and documented, the interpretation of the weight of evidence from laboratory, mesocosm, and field data should be clear and defensible.

4.5 Criteria for Selection of Mesocosm System

The study team recommends that mesocosms be used as the starting point for the General Test Scheme. Criteria for selection of mesocosm systems are presented below. There are many mesocosm designs available with a range of advantages and disadvantages (Hoffmann, 1993; Swift et al., 1993).

One particular type of mesocosm, commonly referred to as "artificial streams," has been employed in a variety of situations. They consist of quite large and robust self-contained units that are placed either indoors or outdoors. Although the advantages of community-level studies of this type are greater ecological complexity and a higher level of environmental realism, the use of large-scale mesocosms also present considerable logistic and financial problems that may limit the number of treatments and the degree of replication that can be employed in any experiment. Specific shortcomings of these systems are:

- Excessive cost limits the economic feasibility of replications;
- Require large areas;
- Are usually site-specific (too bulky to be transported elsewhere);
- Are usually quite inflexible such that the original design cannot be altered to best suit the changing needs of a particular study; and
- Are plagued by the so-called "pseudo-replication syndrome" where multiple channels are used to increase the number of treatments rather than replicating a smaller number of treatments. (Guckert, 1993).

Although the study team is not currently advocating a particular design, it cautions against the use of large-scale mesocosms for the reasons listed above. The ultimate choice of mesocosm design will depend upon an evaluation of the strengths and weaknesses of each candidate system versus the criteria listed below.

Because mesocosms will play a major role in testing the conceptual model, the study team recommends the use of more flexible systems that can be deployed in a variety of sites to allow for extensive testing of the conceptual model.

A number of important criteria have been identified in association with the successful use of mesocosms in an integrated approach to environmental risk assessment in rivers (Touart, 1994). These criteria include aspects of the system that reflect environmental realism such as:

- The ability to represent the flow-through nature of the system;
- The ability to generate uniform riffle velocity throughout the mesocosm;

- The ability to simulate natural temperature, water quality and light regimes;
- The capacity to house periphyton and benthic invertebrate communities and, potentially, fish;
- The ability to manipulate the system with respect to ambient environmental variables including water velocity, chemical parameters, habitat structure (e.g., substrate composition, etc.); and
- The capacity for an experimental duration of at least four weeks.

Other criteria relevant to the collection and interpretation of data generated from the mesocosm include:

- A practical and easily maintained dosing mechanism for the system;
- The capacity for sufficient replication enabling rigorous statistical testing;
- The capacity to accommodate subsampling within each treatment replicate;
- The proven ability of the system to be used to address the types of questions under study; and
- A system design that promotes relative ease of use (i.e., modular, transportable, easily cleaned, etc.).

Each candidate system should be "scored" against the above criteria and the mesocosm design that receives the highest score should be selected.

Scaling up from mesocosms to field situations will be facilitated by the use of transportable mesocosm systems that can be placed along the stream in the same area as field monitoring takes place. Scaling down from the field to mesocosms can yield more defensible data if the mesocosm incorporates some of the critical flow velocity, water quality, temperature, light and substrate conditions of the stream. Again, a transportable mesocosm would aid the scale down/scale up process, because detailed measurements of *in situ* conditions can then be used to adjust test conditions in the mesocosm "in real time." Field monitoring conducted concurrently with a mesocosm experiment will help address temporal scaling issues, because data will be generated during the same season. The longer the mesocosm runs, the more temporal factors can be accounted for. However, tradeoffs between longer-term experiments and artifacts caused by the smaller-scale communities will start to occur the longer the experiment runs.

4.6 Detailed Experimental Testing Scheme

The detailed testing scheme illustrates the specific steps required in order for the complete experimental testing of the conceptual model (Figure 4-4). The central section of Figure 4-4 with bold outline is the main flow-path of the test scheme. This detailed test scheme does not include field design.

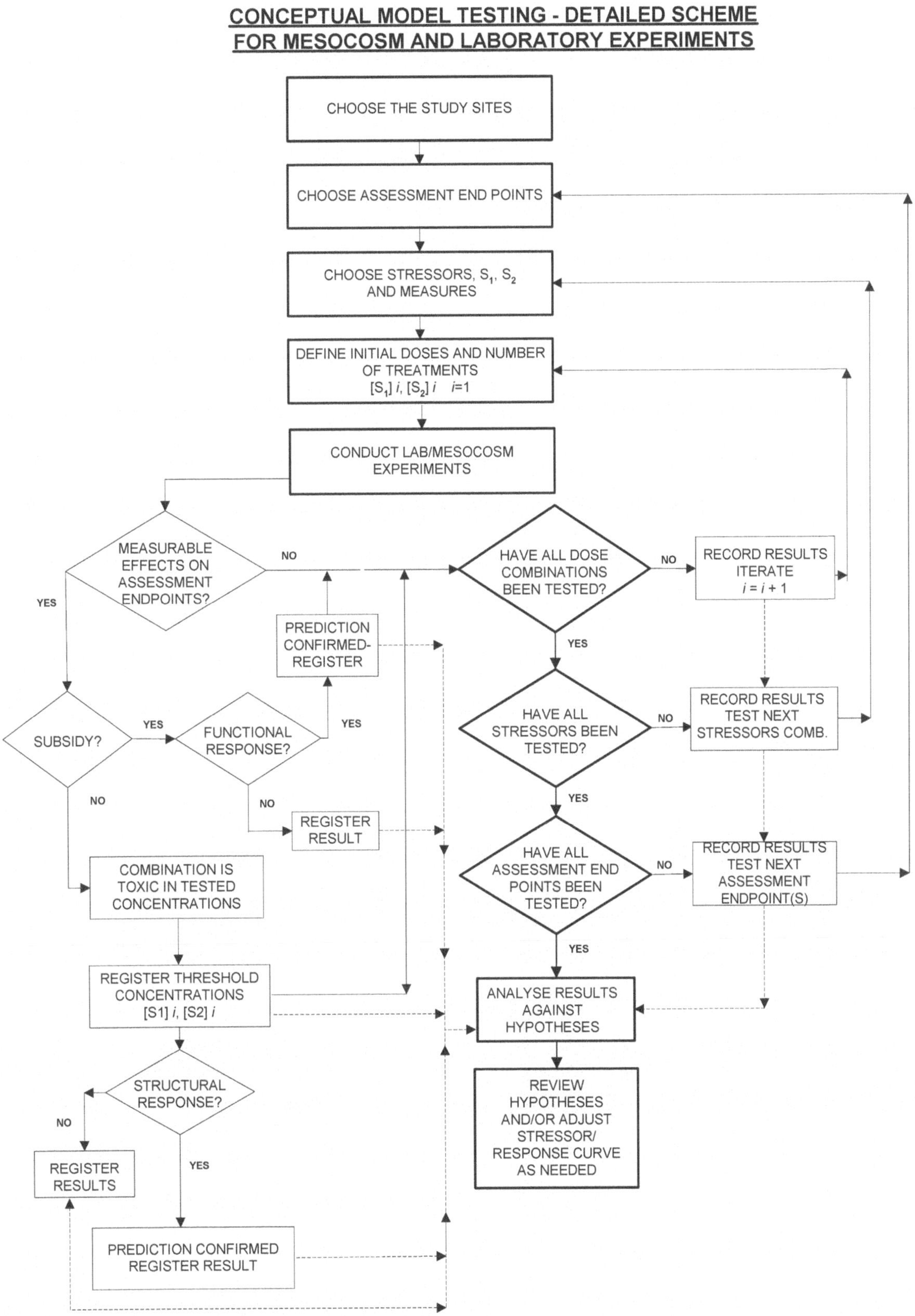

Figure 4-4. Conceptual Model Testingó Detailed Scheme.

The initial steps of the detailed experimental test scheme include selection of assessment endpoints, choosing the study site, selection of stressors and endpoints, and the determination of doses and number of treatments. Each of these steps is described in the following sections. Once the experiments have been designed, the result of each test is interpreted in relation to the three risk hypotheses. Further steps depend on whether the risk hypothesis predictions have been confirmed. The ultimate product of the detailed test scheme is a refined conceptual model.

4.7 Assessment Endpoints

Assessment endpoints are the environmental values—the ecological entities and the desired attributes of those entities—to be protected. To be useful, they need to be ecologically relevant, susceptible to the stressors of concern, and relevant to risk management goals (U.S. EPA, 1998). Ecologically relevant endpoints reflect important characteristics of the system, and may occur at any level of organization—individual, population, community, ecosystem, or landscape.

A useful assessment endpoint is specific, easily measured, ecologically, scientifically or socially significant and likely to produce results that are easy to interpret. Derivation of assessment endpoints often begins with statements of broad goals such as "an unchanged benthic community" or "no significant loss of fish production." These goals, which define the most sensitive, exposed, or important ecosystem components, are useful starting points but should be refined to more specific endpoints. Damage to the benthos, for example, might be assessed by monitoring gross community characteristics (species richness, population density, diversity) or by monitoring a sensitive stonefly species. Similarly, assessment endpoints for a fish community could include maintenance of species association similar to that in unaffected upstream reaches, or it could be based on populations of one or a few signal species such as salmonid (highly sensitive) or pike (top predator).

In the mesocosm and field experiments, the choice of assessment endpoints will vary depending on the particular stressors under consideration. Selection of endpoints must be based on the conceptual model and how it is predicted to apply to the stressor(s) being used. Selections of endpoints take into account sensitivity to specific stressors, both at the laboratory level (e.g., via toxicity testing) and at the mesocosm level.

The mesocosms will contain only microbes, algae, and benthic invertebrates, along with physical habitat, so endpoints must be chosen from those groups. More general ecosystem-level endpoints may also be appropriate, for example community respiration, organic matter removal or nutrient retention. Section 4.9 discusses the specific choices of endpoint available in each of these trophic levels.

Mesocosm experiments study the effect of a combination of stressors at a community-level, but only a limited number of "treatments" are economically viable. This is why mesocosm studies draw upon the findings of screening laboratory tests that examine dose-response curves obtained by testing individual stressors and responses for a chosen endpoint, as follows:

- Initial dose-response testing for individual stressors (single-species study)
- Dose-response testing for a variety of stressor combinations (single-species study)

The latter will then be used as a guideline for the range of stressor concentrations to be used in the mesocosm studies.

The assessment endpoints to be monitored in each experiment must be carefully chosen based on the conceptual model and how it is predicted to apply to the stressor(s) being used. The decision of which ecosystem attributes to measure should incorporate both the mode of action and transport of the stressor and the aspect of the general model to be tested. For example, an experiment to determine whether low levels of sedimentation are stimulatory (as the conceptual model would predict) would probably use community respiration or algal production and community structure as endpoints because these are the ecosystem components most likely to demonstrate the effect, if it exists. On the other hand, a more conventional toxicant such as zinc might be assessed through its effect on benthic invertebrates, which would probably evince toxic effects soonest and most clearly. These effects would be investigated for both tolerant and non-tolerant species.

When considering the routes of stressor effect as a preface to choosing the appropriate endpoint, it is vital to consider the possibility of indirect effects. Ecological theory and a growing body of experimental evidence suggest that indirect effects, those with one or more intermediary species between the stressor and the ultimate effect, may be at least as important as direct toxicity or stimulation. Many indirect effects are mediated through food chains. For example, a change in the light regime in a stream can increase algal production, a predictable direct effect. However, the augmented algal production often leads to both increased productions of benthic invertebrates and a shift in species composition toward members of the scraper functional feeding group. Higher benthos production, in turn, may lead to faster growth of fish. It is therefore important to understand the food chain as thoroughly as possible before selecting the endpoints for a particular stressor experiment. Failure to comprehend potential indirect effects has been responsible not only for unexpected results of ecological experiments, but also some of the most spectacular failures of fishery management and ecosystem manipulation.

4.8 Site Selection

4.8.1 Site Attributes

The attributes of lotic sites needed to conduct experimental research comprise a critical consideration in any research effort, including multiple-stressor research. The site attribute list presented below is based on the assumption that field-based mesocosm experiments, coupled with field observations, would yield the most reliable information on effects of multiple stressors. The following criteria for site selection are proposed:

Environmental Criteria

- Stream(s) of comparable size/flow, that are representative of those typically receiving discharges in the region.
- Access to riffle communities of organisms for use in mesocosms or as part of observational studies.
- Presence of a robust biological community that is not already significantly affected by aquatic stressors such that diversity or productivity is impaired.

- ♦ Moderate to low flow variability that would not compromise field or mesocosm experiments conducted in the river or in stream-side systems.

Logistical Criteria

- ♦ Access to an existing water_supply system (e.g., intake pumps) that could deliver sufficient quantities of water to stream-side mesocosms.
- ♦ Presence of a WWTP discharge, allowing for 1) taking of water from upstream of the discharge, 2) discharge of experimental wastewater to a treatment plant, and 3) access to monitoring data from upstream and downstream of the discharge to provide ancillary data.
- ♦ Access to the site for at least two seasons per year spanning the growing season.
- ♦ Potential for long-term experimentation (e.g., up to three years in study duration) with minimal disruption to mesocosms or field sites.
- ♦ Good potential to obtain necessary permits from state/provincial agencies for both water withdrawal and discharge and for collecting mesocosm biota.

Collaborator Criteria

- ♦ Proximity to laboratory facilities for sample storage and analysis and, ideally, housing for team members, as might be provided by a biological field station.
- ♦ Other concurrent research activity such as ongoing studies by universities, government, or industry that contribute context to multiple-stressor studies.
- ♦ Availability of historical data in the published literature, or in report form, on the biota and major stressors in the drainage system (at a minimum) and ideally on physical and chemical characteristics sufficient for retrospective analysis.
- ♦ Potential for collaboration with willing participants from the institutions, agencies, or industries that have previous experience with the site.
- ♦ Good security for field equipment left at the site and safe working conditions for project field personnel.

4.8.2 Candidate Sites

Sites should be chosen that would allow for a comparison of the influx of stressors in varying concentrations and dilution factors. Data obtained via the U.S. EPA's 1996 Needs Survey based on approximately 12,000 WWTPs indicate that approximately 40% of WWTPs have less than 10% dilution of their effluent into the receiving streams (Figure 4-5). Choosing candidate sites should take into consideration the dilution factors for point-source effluents on the receiving stream to be tested. Effects from low-dilution situations should be compared with receiving environments in which the dilution factors are significantly higher. This may yield some useful information pertaining to stressor effects in relation to dilution.

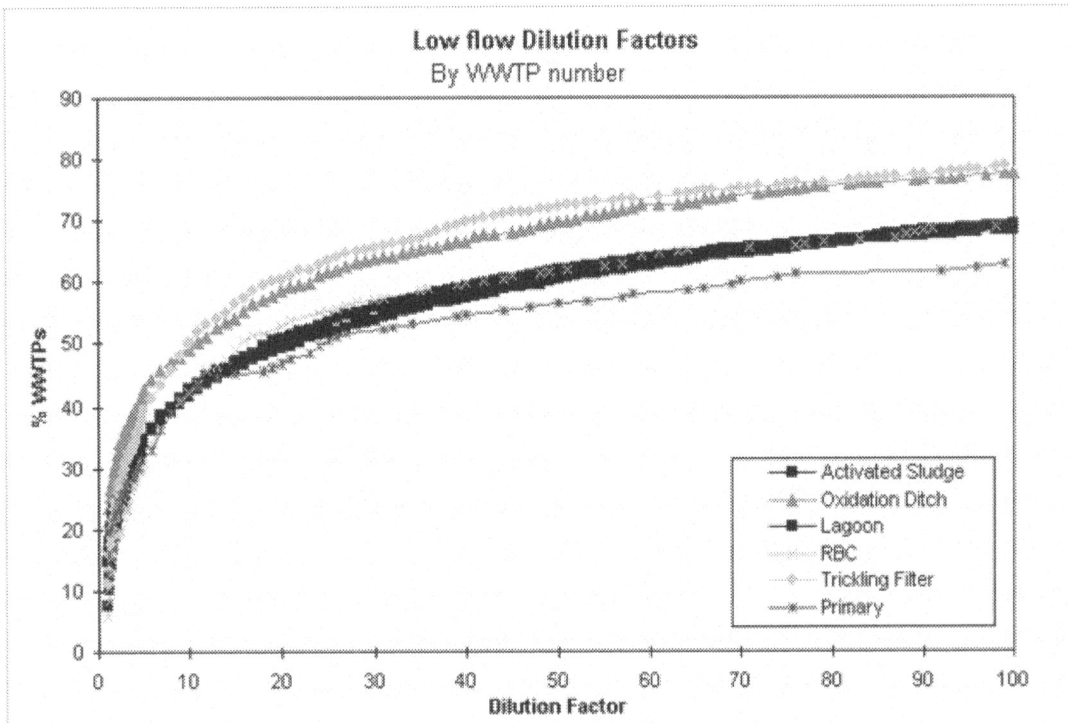

Figure 4-5. Low Flow Dilution Factors by WWTP Numbers. Source: U.S. EPA RF1 dataset.

Several candidate sites for the experimental research phase of the project have been tentatively identified, but other candidate sites that meet the criteria stated in this report could be considered. The three identified candidate sites are the Willamette River in Oregon, the Little Miami River in Ohio, and the Wapiti River in Alberta, Canada.

Several critical attributes were considered when choosing the candidate sites, namely environmental and logistical criteria, including excellent background information, substantial information about stressors and stressor effects, and potential access to a WWTP for discharge of experimental effluent. The Wapiti River had the added benefit of hosting ongoing mesocosm research on aquatic stressors, whereas the Little Ohio River has extensive data from several studies conducted at the Experimental Stream Facility (ESF) on the East Fork of the Little Miami River, Ohio (Belanger et al., 1995; Guckert et al., 1996; Belanger, 1997; Belanger et al., 2000, among others).

The study team proposes that, in terms of size and flow, candidate rivers should not be too small, but rather, be representative of a region. In addition, the candidate rivers should not be too large as they likely contain a multitude of multiple stressors, which would add far too many confounding variables. However, because the conceptual model calls for testing in various sites, it is desirable that candidate sites be comparable in terms of size/flows and to a lesser extent, of stream order.

In the selection of the three candidate sites, stream order was comparable at all three sites (all were of 6th to 7th order) but size and flows were less related (Table 4-14).

Table 4-14. Candidate Rivers Compared by Stream Order, Size, and Flows.

Candidate Rivers	Stream Order	Drainage Area (km^2)	Lowest Mean Monthly Flows (m^3/s)
Wapiti River near Grande Prairie, AB	6-7	11,300	12.8 (Feb.)
Little Miami River at Mildford, OH	6-7	3,116	593 (Oct.)
Willamette River at Harrisburg, OR	6-7	8,857	8,111 (Aug.)

It should be observed that site selection is the last issue to be dealt with. First and foremost, one should go through the risk assessment paradigm, selecting the stressors of concern, the endpoints of concern, and, only then, choosing a suitable study site.

Willamette River

- Considerable current research by universities (e.g., Oregon State University) and government (e.g., U.S. EPA), state agencies (e.g., Oregon Department of Environmental Quality), and the National Council for Air and Stream Improvement (NCASI).
- Extensive data base on biological, physical, and chemical attributes.
- Pulp mill on Willamette River mainstem, or on a major tributary (Mackenzie River), that might be willing to receive mesocosm wastewater into their treatment system.
- Diverse biota in tributaries although the mainstem has been affected by various anthropogenic perturbations including channelization and effluents.

Little Miami River

- Extensive database on biological, physical, and chemical characteristic, in particular the food web with stable isotope analysis.
- Several stream mesocosm experiments conducted using surfactants.
- Ecotoxicological threshold tested for periphyton, protozoan, and invertebrate population.
- Studies to assess the robustness of dose-response investigations without replications.
- Possibility that wastewater would be accepted by County water treatment plants.
- Focus of considerable restoration effort over the past two decades.

Wapiti River

- Extensive data base from federal and provincial governments and industry on biological, physical, and chemical attributes.
- Sufficient-sized river, accessible riffle biota, pristine reference areas readily available.
- Source of treated effluents (municipal sewage, pulp mill) with potential for waste shipment to existing treatment facilities.
- Ongoing research on aquatic stressors using mesocosms with ability to leave systems in place for several years; concurrent research by University of Alberta and Environment Canada, National Water Research Institute.
- Logistics are well known because of team participation in mesocosm research. For example, study sites that have ample water, electrical power, and security are already in place. Discharge licenses have been obtained.

4.9 Stressor Selection

4.9.1 Stressor Attributes

Aquatic stressors can coarsely be separated into biological, physical, and chemical stressors. Biological stressors include such things as exotic species, parasites, and pathogens. Physical stressors often include human modifications to in-stream, riparian, or watershed conditions. Within river systems these changes can alter habitat structure, flow, sediment composition, light, DO, or temperature. Chemical stressors probably are the broadest class of aquatic stressors, and include excess nutrients, metals, pesticides, organic carbon, and many other compounds that enter aquatic systems from human sources. The intersection of these three classes of stressors can pose particular problems for aquatic biota because of the diverse modes of action that they represent. This study design will concentrate on chemical and physical stressors because of their pervasive impact in North American rivers. Preferred attributes of candidate stressors are listed below:

Stressor Importance

- Extensive spatial extent—common stressor found in rivers across North America originating either from point or nonpoint sources.
- Extensive temporal dimension—can be found in river systems over long time frames because of persistence or continuous loading.
- Availability of background data, including concentrations in natural systems and acute and chronic toxicities in standard bioassays.
- Has known or suspected acute or chronic impacts on both periphyton and benthic invertebrates.
- Has the potential to interact with other aquatic stressors because of source, mode-of-action, or physical 'mixing' with other stressors.

Stressor Manipulation and Measurement

- Can be used in experimental systems with reasonable precautions.
- Dose can be readily manipulated to span the range from subsidy to toxicant.
- Stressor is well characterized chemically or physically.
- Can operate independently or in combination with other stressors.
- Analytical methods are well developed and costs are reasonable.
- Measurable at concentrations used in experiments—no nominal concentrations.
- Experimental wastewater can be discharged without undue complications.

Societal Considerations

- Stressor has sociopolitical relevance in North America such that society is concerned about potential adverse impacts.
- Stressor is present in particular region(s) where we intend to conduct experiments or where a change in stressor input is under consideration.

Candidate Stressors

Mesocosm experiments will use chemical and physical stressors of environmental importance in North America. Example candidate stressors are listed below. The initial experimental design will be based on a factorial ANOVA with two factors, such as a chemical stressor and a physical stressor. It is recommended that three sequential experiments be conducted using 1) two different toxicants, 2) two different subsidies, and 3) a toxicant plus a subsidy. These treatments may involve different concentrations of the same stressor, as some stressors may be subsidies at low concentrations but convert to toxicants at higher concentrations (as per the hypotheses outlined in Chapter 3.0). Therefore, the same experiment may encompass one, two, or all three options presented above, which will allow us to test other stressors in subsequent experiments. The selected stressors will not necessarily all be chemicals. Rather, a combination of physical and chemical stressors commonly present in the real world will be selected. Some stressors to be considered for use in experiments are:

Chemical stressors include metals such as copper, nickel, or zinc; nutrients such as phosphorus or nitrogen; dissolved organic carbon; surfactants such as alkyl sulphate; and herbicides such as atrazine, 2,4-D, or glyphosate. Physical stressors include fine sediment (i.e., sediment that causes increased embeddedness), elevated temperature, flow variability, and ultraviolet radiation

4.9.2 Selection of Measures

The endpoints to be measured in the multiple-stressor experiments derive from the conceptual model being tested and the structure of the artificial ecosystems being assembled in the mesocosms. For simplicity's sake, the mesocosms will have three trophic levels: primary producers (benthic algae), primary consumers (benthic macroinvertebrates), and decomposers (benthic microorganisms). Endpoints should be selected from each of these trophic levels.

Fish will be excluded from the mesocosms because of the difficulties in maintaining sufficient numbers and diversity to observe stress responses. However, secondary consumers may still be represented in the mesocosms by predatory invertebrates such as leeches and stoneflies.

Most of the endpoints for the multiple-stressor experiments can be traditional measurements such as growth rate, standing crop, or community composition. Methods for assessing benthic algae and invertebrates are well developed, but the microbial community should not be neglected. Microbes are chiefly responsible for most of the ecosystem-level functions such as nutrient cycling.

Suggested endpoints for invertebrates are:

- ♦ Standing biomass;
- ♦ Species richness, evenness;
- ♦ Population densities (individuals/m^2);
- ♦ Composition of functional feeding groups (Merritt and Cummins, 1996);
- ♦ Similarity indices;
- ♦ EPT/Diptera rations; and
- ♦ Emergence rate.

Identifications to the lowest practical taxonomic level (the level that is routinely obtainable without recourse to specialists or advanced techniques such as mounting mouthparts) should be sufficient for these experiments.

For the benthic algae, species composition, chlorophyll *a* concentration biomass, primary production (measured as CO_2/O_2 production), and dominant growth form or algal group (diatoms, green filamentous, cyanobacteria, etc.) are the parallel structural and functional endpoints.

Many of these metrics were developed for field studies (e.g., species richness, % EPT) and are readily transferable to the larger mesocosm designs that can accommodate periphyton or invertebrate communities. However, in order to reliably discover subtle disturbance effects, individual taxa must be examined, and consideration be made in regards to their ecology and sensitivity. The use of single summary indices that lump together meaningful metrics should be avoided, as they are misleading, lack scientific justification, and do not represent complex entities such as whole ecosystems (refer to Section 4.3)

Population densities of bacteria and fungi can be measured by traditional methods such as dilution plates, agar films and the Most Probable Number technique, or indirectly as concentrations of muramic acid (a cell wall component) or phospholipid ester-linked fatty acids (Findlay and Dobbs, 1993). Direct counts facilitated by the addition of fluorescent stains are considered the most accurate measures of total bacterial density (Ward and Johnson, 1996) but include many cells that are nonvital or metabolically inactive.

Functional measurements, such as uptake rates of tritiated leucine into bacterial protein, provide estimates of microbial productivity. The value of all these measures may be challenged given that they can change so quickly. Microbial populations may increase by many orders of magnitude in hours or days when conditions improve or a labile carbon source arrives. It may be better to measure the microbial component in terms of genetic diversity and the range of enzymes they produce.

Endpoints for the microbial component overlap with those for the whole ecosystem because so many processes are microbially mediated. These include respiration rates, decomposition rates, rates of chemical transformations (e.g., detoxification, nitrification of ammonia), and nutrient retention or spiraling length. Depending on the stressors used, other endpoints may be appropriate that measure an effect specific to a particular stressor.

4.10 Test Design and Number of Treatments

The mesocosm experiments must test the hypothesis of additivity between multiple stressors applied to lotic ecosystems. There are two general approaches. The experiment may use a large number of levels or concentrations of the stressors, and few replicates of each (i.e., a regression approach), or only a few levels of each stressor and commensurately more replicates of each (an ANOVA approach). The first approach is appropriate for tests on single stressors in which the shape of the response curve is to be verified. The second approach, however, is more robust for testing the effects of stressor combinations.

Ultimately, the number of treatments is a compromise between the amount of information gained (i.e., the confidence in the conclusions) and the budget. In the first phase of the experimental work, selected stressors will be applied individually to mesocosm communities to document the nature and degree of response to different stressor levels. For these experiments, the stressor levels should be chosen to span the natural range of levels likely to occur. A minimum of four, and preferably five, levels of the stressor are necessary to properly define the shape of the response curve. However, as this information will not be used for hypothesis testing, a relatively small number of replicates may suffice three or even two at each stressor level. Hence, 15 or 16 experimental units are required.

Once the single-stressor experiments have established the shape of the response curves for each selected stressor under mesocosm conditions, appropriate combinations can be selected for testing in combination. If the experiments to establish response curves have been successful, it should be possible to select levels of each stressor that are clearly in the stimulatory (subsidy) or inhibitory (toxicant) range for each stressor. Stressor levels that are close to a cusp between stimulation and inhibition or that are otherwise at the extremes of the natural distribution should be avoided, as they are more likely to produce ambiguous results.

It is suggested that the experiments to test effects of multiple stressors would use a full two-way ANOVA in which every level of Stressor A is tested at every level of Stressor B, as illustrated in Table 4-15. However, this design is not the only design that could be used; for example, a factorial design might be more efficient than a full two-way ANOVA.

Table 4-15. Full Two-Way Analysis of Variance.

		Stressor A		
Stressor B		**Low**	**Medium**	**High**
	Low	1: Low/Low	4: Medium/Low	7: High/Low
	Medium	2: Low/Medium	5: Medium/Medium	8: High/Medium
	High	3: Low/High	6: Medium/High	9: High/High

The design, if replication is included, permits tests of the main effects of Stressor A and Stressor B, and also the interaction between them. The last test is pivotal. If the hypothesis of additivity is correct, the F value (amount of difference between results) for the interaction effect will be nonsignificant. Any departure from strict additivity, either synergistic (more than additive) or competitive (less than additive), will lead to a significant F for the interaction effect in the ANOVA. Inevitably, the statistical analysis of the mesocosm data will be more complicated because of the imposition of time effects, which must be allowed for by blocking or use of a repeated measures design.

As explained above, the 3×3 two-way ANOVA will test for additivity (or the lack thereof). The experimental work will be completed in two phases. In the first phase, selected stressors will be applied *individually* to mesocosm communities to document the nature and degree of response to different stressor levels. The range and numbers of levels to be used are discussed in this section.

Then, once the single-stressor experiments have established the shape of the response curves for each selected stressor under mesocosm conditions, appropriate combinations can be selected for testing in combination under the 3×3 two-way ANOVA model. The example illustrates a 3×3 design, which is recommended for the experiments because a reasonable number of replicates can then be allocated to each treatment, improving the prospects for discovering unambiguous effects. The same design can, of course be expanded to four, five or more levels of each stressor, but such designs would require either a proportionate increase in the budget and accompanying logistical complications, or a restrictive decline in the number of replicates of each combination. The mesocosm experiments would be the first such attempt to establish the additivity of multiple-stressor effects. Consequently, it is more important that the experiments allow a robust and general test of this hypothesis than to describe the exact nature of the interaction. It is best, therefore, to use only a few levels of each stressor and look for a significant interaction effect. More complex experiments to explore the limits and exceptions to the model can wait for further research.

The choice of a 3×3 experimental design, rather than a 2×2 design, follows from the proposed shape of the general response curve for subsidies/toxicants. Ideally, the low level of each stressor would be either zero or the natural or background level in undisturbed ecosystems. The medium concentration would lie within the subsidy portion of the curve, whereas the high level would lie in the toxicant part of the curve. If both stressors conform to the hypothetical curve, then the 3×3 design captures all possible stressor combinations. If the stressor is a toxicant for which the subsidy stage is small or absent, it may be necessary instead to use different concentrations or intensities of the stressor, all having a negative effect. A 3×3 design nevertheless allows more complex interactions between stressors to be discerned than can be done with a simple 2×2 design.

The experimental design presented above tests the hypothesis of additivity between Stressor A and Stressor B because interaction between stressors is almost always present (but not necessarily significant), meaning that effects are not additive. It is the assumption of this method of ANOVA that a given level of Stressor A and a given level of Stressor B each contribute a certain amount of the effect and that these two contributions add their effects without influencing each other. Departures from this assumption are measured by the *interaction* term. Thus, interaction represents the dependence of the effect of say Stressor A on the level of Stressor B. The (sum of squares for) interaction measures the departure of the subgroup means (e.g., group 1: Low/Low in Table 4-15) from the values expected on the basis of additive combinations of the Stressor A and Stressor B means. Any given combination of levels of stressors such as Low Stressor A/Low Stressor B in Table 4-15 may result in a positive or negative deviation from the expected value based on the means for Low Stressor A (i.e., groups 1– 3 in Table 4-15) and Low Stressor B (i.e., groups 1, 4, and 7 in Table 4-15). A large positive deviation of this sort is known as *synergism*. If the combination of levels of the two stressors inhibit each other's effect, this deviation is called *interference* and is often termed *competitive* interaction. Synergism and interference will tend to magnify the interaction sum of squares. Hence, testing for interaction is a key procedure for our proposed ANOVA approach and, in fact, is an important and standard procedure in any two-way ANOVA.

It is stated in Section 4.9 that it is more important that the experiments allow a robust and general test of the hypothesis than to describe the exact nature of the interaction. Because the study team also recommend using only a few levels of each to meet this objective, a simple method for revealing the nature of a significant interaction is to inspect the means of a summary data table, which could be laid out as in Table 4-15. A pattern of relative magnitudes representing no interaction, but following the expected pattern of responses as described in Section 4.9, is shown in Table 4-16. Experiments yielding interaction between stressors could produce the pattern of relative magnitudes shown in Table 4-17.

When the pattern of signs expressing relative magnitudes is not uniform, as in the latter example, interaction is indicated. It is important to realize, however, that interaction is often present without the change in the *direction* of the differences since only the relative magnitudes may be affected. Regardless, the statistical test performed under ANOVA will test whether the deviations are larger than can be expected from chance alone.

Table 4-16. Example: Full Two-Way Analysis of Variance Showing a Pattern of Relative Magnitudes Representing No Interaction.

				Stressor A		
		Low		Medium		High
	Low		<		>	
Stressor B	Medium		<		>	
	High		<		>	

Table 4-17. Example: Full Two-Way Analysis of Variance Showing a Pattern of Relative Magnitudes Representing Interaction between Stressors.

				Stressor A		
		Low		Medium		High
Stressor B	Low		<		>	
	Medium		<		<	
	High		<		<	

Replication in the present context refers to whole mesocosms, and therefore is not a minor decision, given the cost and effort attached to each unit. For the experiments using the 3 × 3 design described above, four or five replicates in each cell are recommended. Three replicates per cell are the minimum that will permit a meaningful measure of cell variance, which the statistical method needs to compute interactions. However, given the natural variability to which model ecosystems are prone, a higher level of replication is strongly recommended. Given nine cells in the treatment matrix, four replicates per cell require that 36 mesocosms be constructed, and maintained, and monitored. Five replicates (45 mesocosms) would produce a further improvement in sensitivity compared with the three-replicate minimum, and therefore might be worthwhile if responses are variable and the budget is flexible. More than six replicates per cell are unlikely to yield a further improvement sufficient to justify the added expense and complexity. Typically, plots of sample variance against replication reach an asymptote near six replicates.

Conversely, the ANOVA could still proceed with three or even two replicates per cell if budget restrictions require it. Although two replicates cannot provide a meaningful variance estimate for each cell, it is sufficient to allow interaction terms to be computed, and the duplicate mesocosms would at least provide some indication of the range of responses likely to be encountered.

The experimental work will be completed in two phases. In the first phase, selected stressors will be applied individually to mesocosm communities to document the nature and degree of response to different stressor levels. The range and numbers of levels to be used are discussed in Section 4.10. Then, once the single-stressor experiments have established the shape of the response curves for each selected stressor under mesocosm conditions, appropriate combinations can be selected for testing in combination under the 3×3 two-way ANOVA model.

4.11 Quality Assurance/Quality Control

The quality assurance/quality control program for the multiple-stressor experiments must meet the standards for a large, multiperson research program. Standard elements of such a program are 1) assigned responsibilities of each participant; 2) consistent documentation of activities and analyses; 3) data quality checks at each stage of analysis and compilation; and 4) archiving of samples. Because of the size of the anticipated experiments, and the large number of data that will be produced, a formal and detailed quality assurance and control program will be necessary.

4.11.1 Responsibilities of Participants

The multiple-stressor experiments will be multidisciplinary, and may involve scientists and technicians from a number of different organizations and specialties. It is imperative in any group undertaking that every participant has a clear understanding of his or her roles and responsibilities. This delineation is necessary to avoid misunderstandings and spontaneous decisions that too often bog down multidisciplinary studies. A single person should be appointed Study Director, with the responsibility for the overall operation of the study. Other participants should have their responsibilities settled and given in writing before any work begins. Lines of communication and forums for exchanging ideas and information (regular meetings, computer chat rooms, etc.) should be established with a flow chart that is also made available to every participant.

It is not the intention of this partitioning of responsibilities to impede collaboration among the group members. On the contrary, cooperation and interchange of ideas are essential to the success of the project. The flow chart ensures that everyone knows how group decisions are made, so that: 1) all members have an opportunity to participate in any decision that affects them; and 2) any changes in procedure (such as a decision to change sampling frequency) are quickly communicated to everyone. Poor communication is one of the most common failings in large research projects.

As part of the partitioning of responsibilities, one member of the team should be appointed as Quality Control Officer for the project. The QC officer would be responsible for establishing and maintaining quality control and assurance procedures, and especially for ensuring that the quality control measures are consistently carried out. It is far more effective to vest this authority in one person than to depend upon the memories of individual scientists who are busy with other things. It is also necessary to give the Quality Control Officer the authority to implement the quality assurance plan.

4.11.2 Documentation of Procedures

Consistent documentation is the essence of a quality assurance program. Every method of collecting, processing and analyzing samples must be written down in a clear and accessible way so that anyone can easily find out how any particular sampling was done. If a published procedure is being used, the documentation must say where the procedure can be found and any modifications used. This standard is essential for accuracy, especially if there is a change of personnel on the project. Similarly, the inevitable changes in procedure that occur in the midst of the project must be documented.

It is also necessary to establish and follow procedures for identifying and tracing individual samples. Standard sample information includes the date and time, the precise location of the sample (e.g., mesocosm replicate, stream sampling site), the name of the collector, and the disposition of the sample (i.e., what it is collected for). All this information must be recorded on data sheets that accompany each batch of samples to the laboratory where analysis is carried out. Each individual sample must bear an unambiguous, permanent label that links the sample to the data sheet. Analytical results should then be recorded in computer files that contain all the information for every sample.

Where samples are transmitted through several hands before the analytical results are obtained (as for instance, samples sent to a commercial laboratory by courier), a documented chain-of-custody procedure is recommended. A chain of custody form accompanies each batch of samples, with a signed copy retained by each person that handles the samples. Lost or damaged samples can then be easily traced to pinpoint where the loss occurred. Information on lost or compromised samples should still be entered into the computer files, so that any data gaps are explained.

Every computer file must be backed up regularly. Failure to follow this simple rule is probably responsible for more catastrophic data losses than any other cause. Because computer disks infrequently fail, there is a strong temptation to forget or put off the tedious business of making back-up copies. It is worthwhile to invest in an automatic back-up system (such as a Zip drive) that will make fresh copies of every file on a regular schedule, usually daily. It is equally important that all computer files be clearly labeled, with descriptive titles, an index of contents and reasonable protection against accident and vandalism.

4.11.3 Data Quality Checks

Data quality checks include measures to ensure laboratory analyses are reliable, and measures to limit the intrusion of errors in computer data files. Quality control procedures for chemical analyses are well established, and consist of duplicate samples, blanks, and internal standards. Field blanks, samples that are treated identically to regular samples except that no new material is collected (for example by filtering distilled water), should be included with every sample batch. Laboratory blanks are clean samples that reveal contamination or inaccuracy in laboratory methods. Similarly, samples can be split or duplicated in the field or the laboratory to monitor the variability between samples or analytical runs. Internal standards are used in the laboratory to ensure that chemical analysis results are consistent and unbiased.

There are well-established rules for quality control in data analysis. As a general rule, a blank and a duplicate should accompany every group of ten samples. Similar checks are necessary for biological procedures such as toxicity tests, enumeration of invertebrates or microorganisms, or measures of process, such as respiration, organic matter decay or nutrient export. The exact nature of these checks will vary according to the procedure used. Two rules are paramount: 1) the quality control checks should be established from the outset, before any experimental work begins; and 2) they should be repeated regularly as a routine part of sampling and analysis. If a sampling or analytical procedure is changed over the course of an experiment or study, the quality control procedures should be adjusted at the same time.

The multiple-stressor experiments generate a large amount of data. Every time that data are transcribed from one form to another, for instance when paper records are entered into the computer, there is an inevitable incidence of errors. The purpose of a data quality check is to minimize the number of errors and to discover errors when they occur.

The quality control officer, in conjunction with other team members, should establish *data quality goals* for the project. These goals set the tolerable limits for errors in the final data set, and are expressed as an error rate. In smaller data sets, those containing fewer than a hundred entries, it may be reasonable to aim for zero errors. For larger data sets, error rates of 0.1% to 1% are typical.

Comparing the computer data file against the original data for a random subset of each file ensures compliance with data quality goals. A person or persons other than those who entered the data should do this, and the check should include at least 10% of the data. Every error should be reported and the error rate calculated. Simple errors such as typographical errors will dominate, and should be repaired when they occur. A field in the computer file should be reserved for notations indicating a correction. More complex errors such as out-of-range results (e.g., a respiration rate less than zero) should be flagged and brought to the attention of the person responsible for that part of the project.

Data entry errors are only one source, and the simplest, of errors in data files. The more difficult errors to locate are suspicious results from laboratory tests and similar analyses. Consequently, it is imperative that someone familiar with each facet of the project carefully reviews all laboratory data as they are received. The reviewer should inspect the quality control data provided by the laboratory (e.g., internal standards, blanks, and duplicates) and also compare the data against the accumulated data on file. An abrupt or unexpected change in a variable (e.g., disappearance of a species or sudden peak in algal growth rate) should be flagged for confirmation from the laboratory or the investigator responsible for collecting the sample.

The key to successfully controlling errors in data is to *investigate questionable entries as they occur.* Too often, scientists only become aware of suspicious datum months after the fact, when they begin earnest analysis of the data. By that time samples have expired or been discarded, experiments have moved on or ended and the seasons have changed, making it impossible to confirm, replace or interpolate the suspicious datum. Data files should be continually checked for data quality as they are assembled and only files for which all known errors are repaired or flagged should be released for analysis.

4.11.4 Sample Archiving

This is the simplest quality control procedure to implement. Wherever possible, samples should be retained until all quality control checks are completed, and preferably until the end of the study. If samples are properly preserved, labeled and organized, it is possible to re-analyze a sample later to confirm the results from the first test. Even samples that have deteriorated may still be useful. A minimum retention time should be established for each kind of sample collected (e.g., benthic invertebrates, algal growth slides, and sediment for respiration) and a discard date entered for each. When samples are destroyed, the date of destruction should be entered on the data sheet for that batch of samples, so investigators do not waste time looking for samples that no longer exist.

4.11.5 External Audits

External audits should be conducted to ensure that samples are being collected and handled according to the written procedures. This involves having an auditor in the field when samples are being collected. The auditor will have the appropriate data collection procedures and will compare what is done in the field with what is written in the procedures. Any deviance from accepted procedures is written up as a "finding," to which the project manager will need to respond through retraining of personnel or a review of the procedure for applicability. External audits can cover the collection of field data, set-up of mesocosms or experiments, data entry practices, chain-of-custody procedures, storage and transportation of samples, audits of the laboratory testing procedures, etc. Action items with responsibilities and completion dates are usually developed from findings. The project manager has the responsibility to ensure that these completion dates are met and that the finding can be resolved.

4.12 Cost Estimate

The iterative approach to the study design will require a preliminary planning program, a laboratory experiment stage, a field mesocosm study, and a field observation phase to corroborate mesocosm results. Follow-up studies will be required to further develop the conceptual approach and method. Preliminary cost estimates (USD) are provided in Table 4-18.

Table 4-18. Preliminary Cost Estimates (USD) for the Study Design.

Design Component	Estimated Cost
Preliminary planning	$40,000
Laboratory experiment	$60,000
Mesocosm study	$180,000
Field observations	$80,000
Follow-up laboratory testing:	$60,000
Follow-up mesocosm study:	$180,000
Estimated Total	**$600,000**

The estimated total cost would be on the order of $600,000 USD, to be divided over a two-year period. These estimates have a high degree of uncertainty, as they are not directly linked to the study location and final design, type of equipment used, etc. More accurate budget estimates would require finalization of a number of research components. Site logistics such as ease of access, convenient waste disposal facilities and proximity to analytical laboratories may alter the costs associated with the study design. Other factors that will affect costs include the type of experiments conducted and mesocosm apparatus used. Some form of blending of funds and in-kind contributions will help to offset some of the costs of the proposed study.

APPENDIX A

LIST OF DATABASES SEARCHED

APILIT. 1965-Oct 2000. API EnCompass, New York, NY.

Aquatic Sciences and Fish Abstracts. 1978-Oct 2000, Cambridge Scientific Abstracts, Bethesda, MD.

Biosis Previews. 1969-Oct 2001. BIOSIS, Philadelphia, PA.

CAB Abstracts. 1972-Sep 2000. CAB International, Wallingford, Oxon, U.K.

Conference Papers Index. 1973-Sep 2000. Cambridge Scientific Abstracts, Bethesda, MD.

Dissertation Abstracts Online. 1861-Jul 2000. UMI/Bell & Howell, Ann Arbor, MI.

Ei Compendex. 1970-Oct 2000. Engineering Information, Inc., Hoboken, NJ.

Energy Science and Technology. 1974-Oct 2000. U.S. Department of Energy, Oak Ridge, TN.

Enviroline. 1975-Oct 2000. Congressional Information Service, Bethesda, MD.

Environmental Bibliography. 1974-Jul 2000. Environmental Studies Institute, Santa Barbara, CA.

FEDRIP: Federal Research in Progress. Oct 2000. National Technical Information Service, Springfield, VA.

FLUIDEX: Fluid Engineering Abstracts. 1972-Sep 2000. Elsevier Science, SPD, New York, NY.

General Science Abstracts/Full-Text. 1984-Sep 2000. The HW Wilson Company, Bronx, NY.

GeoArchive. 1974-Sep 2000. Geosystems, Oxon, U.K.

GEOBASE. 1980-Oct 2000. Elsevier Science, SPD America, New York, NY.

GeoRef. 1785-2000. American Geological Institute, Alexandria, VA.

Inside Conferences. 1993-Oct 2000. The British Library Document Supply Centre, Wetherby, West Yorkshire, U.K.

JICST-Eplus. 1985-2000. Japan Science and Tech Corporation, Information Center for Science and Technology, Tokyo, Japan.

Lewis, W.M., Jr. 1999. Investigations of the relationship between biodiversity and ecosystem function by examination. NSF Dissertation Research, University of Colorado at Boulder, CIRES (Coop. Inst. for Research in Environmental Sciences).

McGraw-Hill Companies Publications Online. 1985-Oct 2000. The McGraw-Hill Companies Publications Online, Inc., Hightstown, NJ.

Meteorological and Geoastrophysical Abstracts. 1970-Nov 2000. American Meteorological Society, Boston, MA.

NTIS: National Technical Information. 1964-Nov 2000. National Technical Information Service, U.S. Department of Commerce, Springfield, VA.

Oceanic Abstracts. 1964-Nov 2000. Cambridge Scientific Abstracts, Bethesda, MD.

Pascal. 1973-Oct 2000. Scientific and Technical Information Institute of the Centre National de la Recherche Scientifique, Vandoeuvre-les-Nancy, France.

Pollution Abstracts. 1970-Nov 2000. Cambridge Scientific Abstracts, Bethesda, MD.

Science. 1996-July 1999. American Association for the Advancement of Science, Washington, D.C.

TULSA (Petroleum Abstracts). 1965-Oct 2000. Petroleum Abstracts, The University of Tulsa, Tulsa, OK.

WasteInfo. 1974-2000. Waste Management Information Bureau, Abingdon, Oxon, U.K.

WATERNET. 1971-Q3 2000. American Water Works Association, Denver, CO.

Water Resources Abstracts. 1967-Sep 2000. Cambridge Scientific Abstracts, Bethesda, MD.

Wilson Applied Science and Technology Abstracts. 1983-Sep 2000. The HW Wilson Co., Bronx, NY.

Wilson Biological & Agricultural Index. 1983-Sep 2000. The HW Wilson Co., Bronx, N.Y.

Glossary of Terms

Additive	Each substance contributes to the same toxic effect in the same receptor by the same mechanism, with the observed toxicity equal to that expected by adding the toxicities of the individual agents.
Antagonism and Synergism	"A consensus is lacking in the literature regarding the operational definitions of synergism and antagonism among multiple stressors" (Folt et al., 1999). We present the definitions associated with three models of the combined effects of multiple stressors by Folt et al. (1999). The literature was reviewed with this understanding of the various models in mind. *Simple Comparative Effects Model* : The effect of stressors in combination is equal to the effect of the single worst or dominant stressor (Bruland et al., 1991 *in* Folt et al., 1999). This is somewhat analogous to Liebig's law of the minimum where only the nutrient that is most limiting affects growth. Similarly, when the worst stressor is present, lesser stressors have no additional impact. *Additive Effects Model*: Synergism or antagonisms occur when the combined effect of multiple stressors is greater (synergism) or less than (antagonism) the sum of effects elicited by individual stressors (Hay et al., 1994 in Folt et al., 1999; Hay 1996 in Folt et al., 1999). This is the most common scientific use of these terms. The additive model has been used frequently on studies of pesticide or metal mixtures. *Multiplicative Effects:* The combined effect of multiple stressors is equal to the product of individual effects. This model has been used to describe a number of competitive interactions and is applicable when stress from one source can be further operated on probabilistically by another source. For example, if low light reduces yield to 55% of optimal and low nitrogen reduces yield to 45% of optimal, combined effects of low light plus low nitrogen would be predicted to be (0.45 x 0.55 ~ 0.25) or 25% of optimal yield (Folt et al., 1999).
Body Burden	The amount of a substance in the body that is either being metabolized, stored or excreted.
Chemical Concentration	Quantifiable amount of a chemical in environmental media.
Chemical of Concern	A chemical in environmental media that is being studied for its potential presence and impact on a receptor.
Chironomids	Members of the dipteran order of insects, chironomids are true flies, having only one set of complete wings.
Competitive Interaction	The combination of levels of the two stressors inhibits each other's effect (e.g. elevated total dissolved solids may reduce metal toxicity to fish because of competition among cations for binding sites on the gill surface).
Conceptual Model	A qualitative or descriptive statement of hypotheses concerning the nature of the potential causal relationships between human activities, the resulting natural and anthropogenic stressors, and their impacts on human and ecological systems (U.S. EPA 1992, 1998). Conceptual models are often explained by using schematics that illustrate causal relationships. Conceptual models generate "impact hypotheses" that guide the collection and interpretation of data (Harwell and Gentile, 2000). The conceptual model can be revised at each stage of an environmental risk assessment.

Contaminant	A chemical in the environmental media that has either been introduced by human activity or concentrated to levels not naturally found in the area, so that adverse effects may be occurring.
Cumulative	The combination of effects from different activities or sources in an area that has confounding effects on an ecosystem that may not be visible at the individual level.
Cumulative Effects	The impact on the environment that results from the incremental impact of a human activity when added to other past, present and reasonably foreseeable future activities (Harwell and Gentile, 2000). This definition could refer to a single stressor (such as BOD) that increases incrementally with each additional discharge point to a river. However, this definition is broad enough to include the action of multiple stressors within the same timeframe and same spatial context, and also across time periods. With this broad definition in mind, it can be concluded that cumulative risk assessments and cumulative impact assessments are two different approaches used to assess cumulative effects.
Dose	A measure of integral exposure. Examples include the amount of a chemical ingested, the amount of a chemical actually taken up and the product of the exposure concentration and the duration of the exposure.
Ecological Endpoint or Assessment Endpoint	An explicit expression of valued environmental attributes that can be defined in operational terms and are amenable to measurement or prediction (Harwell and Gentile, 2000). This definition includes both ecological importance and/or relevance and societal value. Selection of ecological endpoints should result from dialogue between the risk assessor and risk manager during the initial phase of a risk assessment and should reflect stakeholder concerns (Harwell and Gentile, 2000). The term "assessment endpoint" is nearly synonymous with the term "valued ecological component (VEC)" used in the impact assessment literature, although a VEC can also simply be a particular species (which would then be synonymous with "receptor").
Ecologically Significant	The selection of assessment endpoints requires an understanding of what constitutes ecological significance. Harwell and Gentile (2000) suggest that ecological significance relates to determining: --whether a chance that is detected or projected in the ecological system of concern constitutes a change of importance to the structure, function, and/or health of the system and exceeds a variance estimate (i.e., placed into the content of natural variability); and --whether such a change in the ecological system is of sufficient type, intensity, extent and/or duration to be important to society.
Effects	A noticeable change in the receptor beyond normal variability caused by a chemical of concern or other stressor.
Eutrophic	High levels of dissolved nutrients, low oxygen levels during warm weather due to high photosynthetic activity
Exposure	The contact reaction between a chemical and biological system or an organism.
Exposure Concentration	The concentration of a chemical in its transport or carrier medium at the point of contact.
Fate and Transport	The movement and final location of a chemical through different media. May include partitioning, uptake and degradation.
Impacts	A noticeable change in the receptor in reaction to a stressor. See also "Effects."

Indicator	A characteristic of the environment that provides evidence of the occurrence or magnitude of exposure or effects. Formal expressions of the results of measuring an indicator are referred to as measurement endpoints. Abundance, yield, and age/weight ratios are indicators of population production. The presence of female-like characteristics in male fish is an indicator of endocrine system disruption.
Indicator Species	A species that is surveyed or sampled for analysis as a representative of the biotic community, some functional or taxonomic group, or some population that cannot be readily sampled or surveyed.
Interference	The mode of action of one stressor interferes with the mode of action of another stressor, thus reducing the total effect of the combination of stressors.
LC50	Concentration of a stressor at which a lethal response is observed in 50% of the test organisms.
Linkages	The pathways within a conceptual model that connect the stressor with the pathway, receptor and endpoint.
Loadings	The discharges to a system from various different sources. Loadings may be industrial or municipal, spills, chemical application, leaching from groundwater, atmospheric deposition, etc. Loadings are often expressed as the quantity of chemicals discharged (or loaded) to a system.
Lotic	Of, relating to, or living in moving water.
Measures of ecosystem and receptor characteristics	These are environmental variables that are not identified as stressors of concern, but that may also cause the measured effects to change. These characteristics describe attributes of systems and the organisms of concern that may affect the relationship between the measured effects and exposure to the stressors of concern. Included in this category are life-history characteristics of the species of concern, ecosystem characteristics that affect the behavior and location of these species, and attributes of the system that affect the movement, intensity and availability of the stressors. Measures of ecosystem and receptor characteristics also are synonymous with the statistical terms "independent variable" or "covariate" (Fairbrother and Bennett, 2000).
Measure of Effect	Variables that respond to stressors. Synonymous with the statistical terms "dependent variable" or "response variable" (Fairbrother and Bennett, 2000).
Measure of Exposure	Quantification of stressors of concern. Synonymous with the statistical term "independent variable"; also synonymous with "treatment" if the study is experimental (Fairbrother and Bennett, 2000). *Fairbrother and Bennett (2000) note that in risk assessment, the differences between a measure of effect, a measure of exposure and measures of ecosystem and receptor characteristics generally are a matter of perspective and the way the environmental manager frames the particular question. What may be perceived as an attribute of concern and a measure of effect in one system may become a measure of exposure in another system if it also affects another attribute of greater concern. It is more important to state clearly the questions that are being addressed: what environmental attributes are of concern, what potential stressors are being considered, and what additional environmental variables are present and may change the impact of the stressors?*

Mesocosm	A composite physical and biological model of an ecosystem, intermediate in scale between a microcosm and macrocosm, with a level of organization as similar as possible to the natural world. Larger scale mesocosms are often referred to as artificial streams.
Multiple	Two or more.
Multiple Stressors	The combined effects of multiple human influences on populations and ecosystems. These commonly include conventional water pollution, toxic chemicals, land-use change, harvesting and many other qualitatively different disturbances (Barnthouse et al., 2000). Two or more stressors acting simultaneously.
Oligotrophic	A body of water with low nutrient levels.
Pathways	The means through which a chemical or other stressor comes in contact with and affects the receptor. For example, at a site with lead-contaminated soil, the pathways for intake of lead in humans are ingestion (soil on foods, children eating soil), inhalation of dust and direct absorption through the skin.
Periphyton	Attached benthic algae that are sensitive indicators of environmental change in lotic waters.
Receptor	The population, community, organism, organ, or target tissue that is affected by the stressor.
Reference Condition Assessment (RCA)	The reference condition is defined as "the condition that is representative of a group of minimally disturbed sites organized by selected physical, chemical, and biological characteristics."
Ruderal	Growing in rubbish, poor land, or waste.
Species Tolerances	The physiological and ecological tolerances and requirements describe the "hardiness" of a species. The more hardy a species is, the greater its ability to adapt to quickly changing environments (i.e., more tolerant)
Stream Order	A method of organizing streams into hierarchies that uses convergence of equal streams to create a higher order. For example, the uppermost tributaries are labeled as first order streams, and the joining of two first order streams results in the creation of a second order stream. Two second order streams converge to form a third order stream, etc. (Strahler, 1957).
Stress	An external influence on a population, community, organism, organ, or target tissue that causes an affect, either positive or negative.
Stressor	The chemical, physical, biotic, or environmental factor that alters an ecosystem beyond the range of normal variation or from its natural state (Breitburg et al., 1999).
Stressor of Concern	A factor identified by the investigator as being outside the normal range of concentration occurrence, usually because it exceeds background or baseline concentrations, levels or frequency, or because it exceeds water or sediment quality criteria. The stressors of concern are a subset of the total stressors acting upon an ecosystem.

Synergistic	Two or more toxic substances interact such that the toxicity of the mixture is greater than would be expected if the substances were acting additively or independently. For example, people who use both alcohol and tobacco have a much higher risk of some cancers than would be expected if these two products were acting additively.
Tolerance	A range of values from 0 to 10 that indicates a species' ability to withstand organic waste. Species with 0 tolerance values are very intolerant of organic waste, whereas species with a 10 tolerance rating are very tolerant of organic waste. These values are based on numerous well-tested indices, most notably the Hilsenhoff species-level biotic index and the Family Biotic Index (Hilsenhoff, 1982, 1987, 1988).

References

Baird, D.J. and G.A. Burton, Eds. *Ecological Variability: Separating Natural from Anthropogenic Causes of Ecosystem Impairment*. SETAC Press: Pensacola, FL, 2001.

Baker, J.L., A.C. Barefoot, L.E. Beasley, L.A. Burns, P.P. Caulkins, J.E. Clark, R.L. Feulner, J.P. Giesy, R.L. Graney, R.H. Griggs, H.M. Jacoby, D.A. Laskowski, A.F. Maciorowski, E.M. Mihaich, H.P. Nelson Jr., P.R. Parrish, R.E. Siefert, K.R. Solomon, W.H. van der Schalie, Eds. *Aquatic Dialogue Group: Pesticide Risk Assessment and Mitigation*. SETAC Press: Pensacola, FL, 1994.

Barbour, M.T., J. Gerritsen, G.E. Griffith, R. Frydenborg, E. McCaron, J.S. White, and M.L. Bastian. A framework for biological criteria for Florida streams using benthic macroinvertebrates. *J. N. Am. Benthol. Soc.* **1996,** *15*, 185–211.

Belanger, S.E. Literature review and analysis of biological complexity in model stream ecosystems: Influence of size and experimental design. *Ecotoxicol. Environ. Safety* **1997,** *36,* 1–16.

Belanger, S.E., J.B. Guckert, J.W. Bowling, W.M. Begley, D.H. Davidson, E.M. LeBlanc, and D.M. Lee. Responses of aquatic communities to 25-6 alcohol ethoxylate in model stream ecosystems. *Aquat. Toxicol*. **2000**, *48*, 135–150.

Belanger, S.E., E.M. Meiers, and R.G. Bausch. Direct and indirect ecotoxicological effects of alkyl sulfate and alkyl ethoxysulfate on macroinvertebrates in stream mesocosms. *Aquat. Toxicol.* **1995**, *33*, 65–87.

Beyers, D.W. Causal inference in environmental impact studies. *J. N. Am. Benthol. Soc.* 1998, *17*, 367–373.

Beyers, R.J., and H.T. Odum. Ecological microcosms. Springer-Verlag, New York, 1993.

Boyle, T.P., J.F. Fairchild, E.F. Robinson-Wilson, P.S. Haverland, and J.A. Lebo. Ecological restructuring in experimental aquatic mesocosms due to the application of diflubenzuron. *Environ. Toxicol. Chem.* **1996**, *10,* 1806–1814.

Breitburg D., S. Seitzinger, and J. Sanders, Eds. The effects of multiple stressors on freshwater and marine ecosystems. *Limnol. Oceanogr*. **1999**, *44*(3), Part 2. 739–972

Calabrese, E.J., and L.A. Baldwin 1998. Hormesis as a biological hypothesis. In: Toxicological Defence Mechanisms and the Shape of Dose-Response Relationships. *Environ. Health Perspect.* **1998**, *106* (Suppl. 1).

Chapman, P.M. Workshop manual. In: *Analyzing and Interpreting Contaminated Harbour and River Sediment Workshop*. Vancouver, British Columbia, Jan. 27–29, 2003.

Clements, W.H., and P.M. Kiffney. Integrated laboratory and field approach for assessing impacts of heavy metals at the Arkansas River, Colorado. *Environ. Toxicol. Chem.* **1994**, *13*, 397–404.

Clements, W.H., and P.M. Kiffney. Validation of whole effluent toxicity tests: integrated studies using field assessments, microcosms, and mesocosms. In: *Whole Effluent Toxicity Testing,* D.R. Grothe, K.L. Dickson, and D. Reed-Jenkins, Eds.; SETAC Press: Pensacola, FL, 1996; pp. 229–343.

Culp, J.M., and C.L. Podemski. Impacts of contaminants and nutrients in bleached kraft mill effluent on benthic insect and periphyton communities: Assessments using artificial sreams, Athabasca River, 1993 and 1994. Northern River Basin Study. SSC-R71-49/3-92E. 1996, 1–70.

Culp, J.M., R.B. Lowell, and K.J. Cash. Integrating mesocosm experiments with field and laboratory studies to generate weight-of-evidence risk assessments for large rivers. *Environ. Toxicol. Chem.* **2000a**, *19*, 1167–1173.

Culp, J.M., C.L. Podemski, K.J. Cash, and R.B. Lowell. A research strategy for using stream microcosms in ecotoxicology: integrating experiments at different levels of biological organization with field data. *J. Aquat. Ecosyst. Stress Recov.* **2000a,** *7*, 167–176.

Culp, J.M, K.J. Cash, N.E. Glozier, and R.B. Brua. Effects of pulp mill effluent on benthic food webs of the Saint John River, Canada. *Environ. Toxicol. Chem.* **2000b** , *22* (12), 2916–2925

Dickson, K.L., W.T. Waller, J.H. Kennedy, and L.P. Ammann. Assessing the relationship between ambient toxicity and instream biological response. *Environ. Toxicol. Chem.* **1992**, *11,* 1307–1322.

Dickson, K.L., W.T. Waller, J.H. Kennedy, L.P. Ammann, R. Guinn, and T.J. Norberg-King. Relationships between effluent toxicity, ambient toxicity, and receiving system impacts: Trinity River dechlorination case study. In *Whole Effluent Toxicity Testing,* D.R. Grothe, K.L. Dickson, and D. Reed-Jenkins, Eds.; SETAC Press: Pensacola, FL, 1996; pp. 287–308.

Dorn, P.B., J.H. Rodgers, Jr., S.T. Dubey, W.B. Gillespie, Jr., and A.R. Gigueora. Assessing the effects of a $C_{14\text{-}15}$ linear alcohol ethoxylate surfactant in stream microcosms. *Ecotox. Environ. Safety* **1996**, *34*, 196–204.

Dubé, M., D. MacLatchy, J. Culp, G. Gillis, R. Parker, S. Courtenay, and C. Gilman. Utility of mobile, field-based artificial streams for assessing effects of pulp mill effluents on fish in the Canadian environmental effects monitoring (EEM) program. *J Aquatic Ecosys. Stress Recov.* **2002**, *9*, 85–102.

Dyer, S.D., and S.E. Belanger. Determination of the sensitivity of macroinvertebrates in stream mesocosms through field-derived assessments. *Environ. Toxicol. Chem.* **1999**, *18* (12), 2903–2907.

Dyer, S.D., C.E. White-Hull, and G.J. Carr. *Characterizing the Effects of Multiple Stressors at the State Level.* Paper presented at the SETAC Workshop on Modeling and Measuring the Vulnerability of Ecosystems at Regional Scales for Use in Ecological Risk Assessment and Risk Management; Seattle, August, 1998.

Edmondson, W.T., and G.G. Winberg. *A Manual on Methods for the Assessment of Secondary Productivity in Fresh Waters,* International Biological Programme Handbook No. 17; Blackwell Scientific: Oxford, U.K., 1971.

Elliott, M. The derivation and value of ecological quality standards and objectives. *Marine Pollut. Bull.* **1996**, *32*, 762–763.

Environment Canada. *Pulp and Paper Technical Guidance Document for Aquatic Environmental Effects Monitoring.* EEM/1998/1, April **1998**. Environment Canada: Ottawa, ON, Canada, 1998.

Fairbrother, A., and R.S. Bennett. Multivariate statistical applications for addressing multiple stresses in environmental risk assessments. In *Multiple Stressors in Ecological Risk and Impact Assessment: Approaches to Risk Estimation,* S.A. Ferenc and J.A. Foran, Eds.; SETAC Press: Pensacola, FL, 2000; pp. 67–116.

Fairchild, J.F., T.W. LaPoint, J.L. Zajicek, M.K. Nelson, F.J. Dwyer, and P.A. Lovely. Population-, community- and ecosystem-level responses of aquatic mesocosms to pulsed doses of a pyrethroid insecticide. *Environ. Toxicol. Chem.* **1992**, *11*, 115–129.

Findlay, R.H., and F.C. Dobbs, Quantitative description of microbial communities using lipid analysis," In *Handbook of Methods in Aquatic Microbial Ecology*, P.F. Kemp, B.F. Sherr, E.B. Sherr, and J.J. Cole, Eds.; Lewis Publishers: Boca Raton, FL, 1993; pp. 271–284.

Foran, J.A., and S.A. Ferenc, Eds. *Multiple Stressors in Ecological Risk and Impact Assessment*; SETAC Press: Pensacola, FL, 1999.

Foran, J.A., and S.A. Ferenc, Eds. *Multiple Stressors in Ecological Risk and Impact Assessment; Approaches to Risk Estimation.* SETAC Press: Pensacola, FL, 2000.

Forbes, V.E., and T.L. Forbes. *Ecotoxicology in Theory and Practice.* Chapman & Hall: New York, 1994.

Fox, G.A. Practical causal inference for ecoepidemiologists. *J. Toxicol. Environ. Health* **1991**, *33*, 359–379.

Gerritsen, J., B. Jessup, and E. Leppo. *Effects of Multiple Stressors on Aquatic Ecosystems* (WERF 96-IRM-2); Water Environment Research Foundation: Alexandria, VA, 2004.

Gibbons, W.D., and K.R. Munkittrick A sentinel monitoring framework for identifying fish population responses to industrial discharges. *J. Freshwater Ecosys. Health* **1994**, *3*, 227–237.

Gilbertson, M. Advances in forensic toxicology for establishing causality between Great Lakes epizootics and specific persistent toxic chemicals. *Environ. Toxicol. Chem.* **1997**, *16*, 1771–1778.

Goodman, I.A., M.E. Jensen, J.K. Andreason, and M.W. Letourneau. *Characterizing Aquatic System Vulnerability in the Columbia Basin: Applying Hierarchy Theory to Scaled Ecological Assessments.* Paper presented at the SETAC Workshop on Modeling and Measuring the Vulnerability of Ecosystems at Regional Scales for Use in Ecological Risk Assessment and Risk Management; Seattle, August, 1998.

Graney, R.L, J.H. Kennedy, and J.H. Rodgers, Eds. *Aquatic Mesocosm Studies in Ecological Risk Assessment;* Lewis Publishers: Boca Raton, FL, 1994.

Guckert, J.B. Artificial streams in ecotoxicology. *J. N. Am. Benthol. Soc.* **1993**, *12,* 350–356.

Guckert, J.B., D.D. Walker, and S.E. Belanger. Analytical chemistry of stream mesocosms exposed to dodecyl alkyl sulfate reveals rapid biodegradation. *Environ. Toxicol. Chem.* **1996**, *15* (3), 262–269.

Haley, R.K., T.J. Hall, and T.M. Bousquet. Effects of biologically treated bleached-Kraft Mill effluent before and after Mill conversion to increased chlorine dioxide substitution: results of an experimental streams study. *Environ. Toxicol. Chem.* **1995**, *14* (2), 287–298.

Hall, T.J. Personal communication. Letter to Stella Swanson, December 14, 2001.

Hall, T.J., R.K. Haley, and L.E. LaFleur. Effects of biologically treated bleached-Kraft Mill effluent on cold water stream productivity in experimental stream channels. *Environ. Toxicol. Chem.* **1991**, *10* (8), 1051–1060.

Harwell, M.A. and J.H. Gentile. Environmental decision-making for multiple stressors: framework, tools, case studies and prospects. In *Multiple Stressors in Ecological Risk and Impact Assessment: Approaches to Risk Estimation,* S.A. Ferenc and J.A. Foran, Eds.; SETAC Press: Pensacola, FL, 2000; pp. 169–226.

Hauer, F.R., and G.A. Lamberti. *Methods in Stream Ecology;* Academic Press: San Diego, 1996.

Heath, A.G. *Water Pollution and Fish Biology*, 2nd ed.; Lewis Publishers, CRC Press: Boca Raton, FL, 1995.

Hill, A.B. The environment and disease: association or causation. *Proc. R. Soc. Med.* **1965**, *58*, 295–300.

Hill, I.R., F. Heimbach, P. Leewwangn, and P. Matthieson, Eds. *Freshwater Field Tests for Hazard Assessment of Chemicals*; Lewis Publishers: London, 1994.

Hilsenhoff, W.L. 1982. Using a biotic index of to evaluate water quality in streams. *Tech. Bull. Wisc. Dept. Nat. Resour.* **1982**, *132*.

Hilsenhoff, W.L. An improved biotic index of organic stream pollution. *Great Lakes Entomol.* **1987**, *20*: 31–39.

Hilsenhoff, W. L. Rapid field assessment of organic pollution with a family level biotic index. *J.N. Am. Benthol. Soc.***1988**, *7*:65–68.

Hoffmann, J.P. Small-scale artificial streams. *J. N. Am. Benthol. Soc.* **1993**, *12*, 356–359.

Hope, B.K. *A New Diet of Worms: Ecologists, Toxicologists and Regulators*. Paper presented at Society for Risk Analysis, Phoenix, AZ, Dec. 6–9, 1998.

Jones, K.B., K.H. Ritters, J.D. Wickham, R.D. Tankersley Jr., R.V. O'Neill, D.J. Chaloud, E.R. Smith, and A.C. Neale. *An Ecological Assessment of the United States Mid-Atlantic Region.* Paper presented at the SETAC Workshop on Modeling and Measuring the Vulnerability of Ecosystems at Regional Scales for Use in Ecological Risk Assessment and Risk Management; Seattle, August, 1998.

Karr, J.R., K.D. Fausch, P.L. Angermier, P.R. Yant, and I.J. Schlosser. Assessing biological integrity in running waters: a method and its rationale. *Ill. Nat. Hist. Surv.* **1986**, Spec. Publ. 5.

Kersting, K., and P.A. Van Wijngaarden. Effects of a pulsed treatment with the herbicide afalon (active ingredient Linuron) on macrophyte-dominated mesocosms. I. Responses of ecosystem metabolism. *Environ. Toxicol. Chem.* **1999**, *18,* 2859–2865.

Kiffney, P.M., and W.H. Clements. Effects of heavy metals on a macro-invertebrate assemblage from a Rocky Mountain stream in experimental microcosms. *J. N. Am. Benthol. Soc.* **1994**, *13*, 511–523.

Kimball, K.D., and S.A. Levin. Limitations of laboratory bioassays: the need for ecosystem-level testing. *Bioscience* **1985**, *35*, 165–171.

Kline, E.R., A. Rocio, J.H. Figueroa, J.H. Rodgers, Jr., and P.B. Dorn. Effects of a nonionic surfactant (C_{14-15} AE-7) on fish survival, growth and reproduction in the laboratory and in outdoor stream mesocosms. *Environ. Toxicol. Chem.* **1996**. *15*, 997–1002.

La Point, T.W., S.M. Melanson, and M.K. Morris. Relationships among observed metal concentrations, criteria and benthic community structural responses in 15 streams. *J. Water Pollut. Control Fed.* **1984**, *56*, 1030–1038.

Lamberti, G.A. Personal communications with Geetha Ramesh, 2002.

Lamberti, G.A., and A.D. Steinman. Introduction. *J. N. Am. Benthol. Soc.* **1993**, *12*, 314–317.

Landis, W.G., L.R. Bodensteiner, and J.F. McLoughlin. The relative risk rank model for regional-scale risk assessment and potential application in freshwater and terrestrial ecosystems. In *Risk Assessment of Environmental Endpoints. Proceedings of a Workshop, University of Aukland October 28-30,*. Lewis, G.D., N.G. Thom, J.E. Hay and K. Sukhia, Eds., Auckland, School of Environmental and Marine Sciences, University of Auckland, Auckland, New Zealand, 1998; pp. 54–67.

Landis, W.G., A.J. Markiewicz, R.A. Matthews, and G.B. Matthews. A test of the community conditioning hypothesis: persistence of effects in model ecological structures dosed with jet fuel JP-8. *Environ. Toxicol. Chem.* **2000**, *19* (2), 327–336.

Lehtinen, K.J, J. Tana, K. Mattsson, J. Hardig, P. Karlsson, C. Grotell, S. Hemming, C. Engstrom, and J. Hemming. 1993. Ecological impact of pulp mill effluents, Parts I and II. Publ fin Natl. Board Water Environ. Adm Series A 133.

Leland, H.V., S.V. Fend, T.L. Dudley, and J.L. Carter. Effects of copper on species composition of benthic insects in a Sierra Nevada, California, stream. *Freshwater Biol.* **1989**, *21*, 163–180.

Lewis, W.M., Jr. Investigations of the Relationship between Biodiversity and Ecosystem Function by Examination; dissertation research; University of Colorado at Boulder, CIRES (Coop. Inst. for Research in Environmental Sciences), 1999.

Lowell, R.B. Discussion paper on critical effect size guidelines for EEM using benthic invertebrate communities, EEM/1997/8. National Environmental Effects Monitoring Office, Environment Canada: Ottawa, 1997.

Lowell, R.B., J.M. Culp, and F.J. Wrona. Stimulation of increased short-term growth and development of mayflies by pulp mill effluent. *Environ. Toxicol. Chem.* **1995a**, *14* (9), 1529–1541.

Lowell, R.B., J.M. Culp, and F.J. Wrona. Toxicity testing with artificial streams: Effects of differences in current velocity. *Environ. Toxicol. Chem.* **1995b**, *14* (7), 1209–1217.

Lowell, R.B., J.M. Culp, and M.G. Dube. A weight of evidence approach for northern river risk assessment: integrating the effects of multiple stressors. *Environ. Toxicol. Chem.* **2000**, *19*, 1182–1190.

Lowell, R.B., J.M. Culp, F.J. Wrona, and M.L. Bothwell. Effects of pulp mill effluent on benthic freshwater invertebrates: food availability and stimulation of increased growth and development. In *Environmental Fate and Effects of Pulp and Paper Mill Effluents*, M.R. Servos, K.R. Munkittrick, J.H. Carey, and G.J. Van Der Kraak, Eds.; St. Lucie Press: Delray Beach, FL, 1996; pp. 525–532.

Lowell, R.B., and J.M. Culp. Implications of sampling frequency for detecting temporal patterns during environmental effects monitoring. *Water Qual. Res. J. Can.* **2002**, 37 (1), 119-132.

Massachusetts Weight-of-Evidence Workgroup. Draft Report. A Weight-of-Evidence Approach for Evaluating Ecological Risks. November 2, 1995.

Matthews, R.A., W.G. Landis, and G.B. Matthews. The community conditioning hypothesis and its application to environmental toxicology. *Environ. Toxicol. Chem.* **1996**, *15* (4), 597–603.

Merritt, R.W., and K.W. Cummins, Eds. *An Introduction to the Aquatic Insects of North America*, 3rd ed.; Dubuque, IA: Kendall/Hunt Pub., 1996.

Munkittrick, K.R., and D.G. Dixon. A holistic approach to ecosystem health assessment using fish population characteristics. *Hydrobiologia* **1989**, *188/189*, 123–135.

Odum, E.P. Trends expected in stressed ecosystems. *BioScience* **1985**, *35*, 419–422.

Odum, E.P., J.T. Finn, and E.H. Franz. Perturbation theory and the subsidy-stress gradient. *BioScience* **1979**, *29* (4), 349–352.

Parsons, P.A. The *Hormetic Zone, Habitats and Phantom Radiation Risks*. AINSE Occasional Papers, NSW, Australia, 2002.

Paulson, C., S. Reeves, S.P. Canton, T. Moore. *Distinguishing the Relative Influence of Habitat and Water Quality on Aquatic Biota* (WERF 98-WSM-1). Water Environment Research Foundation: Alexandria, VA, 2001.

Perry, J.A., and N.H. Troelstrup Jr. Whole ecosystem manipulation a productive avenue for test system research. *Environ. Toxicol. Chem.* **1988**, *7* (11), 941–951.

Pratt J.R, and J. Cairns, Jr. Ecotoxicology and the redundancy problem: understanding effects on community structure and function. In *Ecotoxicology: A Hierarchical Treatment,* Newman M.C., Jagoe C.H., Eds.; Lewis Publishers: Boca Raton, FL, 1996; pp. 347–370.

Rapport, D.J., and W.G. Whitford. How ecosystems respond to stress. *BioScience* 1999, 49, 193–204.

Rosenberg, D.M., and V.H. Resh, Eds. *Freshwater Biomonitoring and Benthic Invertebrates*; Chapman & Hall: New York, 1993.

Sandberg, R.S., and W.G. Landis. Persistence of the effects of Jet-A in a microcosm with releases from the sediment. *Environ. Toxicol. Chem.* **2001**, *20* (9): 1942–1950.

Schindler, D.W. Detecting ecosystem responses to anthropogenic stress. *Can. J. Fish. Aquat. Sci.* **1987**, *44* (Suppl. 1): 6–25.

Seyle, H. The evolution of the stress concept. *Am Sci.* **1973**, *61*, 692–699.

Shaw, J.L., and J.H. Kennedy. 1996. The use of aquatic field mesocosm studies in risk assessment. *Environ. Toxicol. Chem.* 15: 605–607.

Smith, E.P., and R. Voshell. *MAIS: A Multimetric Benthic Indicator of Ecological Condition for EPA Region III.* Paper presented at the SETAC Workshop on Modeling and Measuring the Vulnerability of Ecosystems at Regional Scales for Use in Ecological Risk Assessment and Risk Management; Seattle, August, 1998.

Stoddard, J.L., A.T. Herlihy, B.H. Hill, R.M. Hughes, C.B. Johnson, P.R. Kaufmann, D.J. Klemm, and F.H. McCormick. *Relationships of In-Stream Chemistry, Physical Habitat and Biological Condition to Land Use in the Mid-Atlantic Highlands Region.* Paper presented at the SETAC Workshop on Modeling and Measuring the Vulnerability of Ecosystems at Regional Scales for Use in Ecological Risk Assessment and Risk Management; Seattle, August, 1998.

Strahler, A.N. Quantitative Analysis of watershed geomorphology. *Am. Geophys. Union Trans.*. **1957,** *38* (1): 913-920.

Suter II, G.W., *Ecological Risk Assessment;* Lewis Publishers: Boca Raton, FL, 1993.

Suter II, G.W. 1996. *Risk Characterization for Ecological Risk Assessment of Contaminated Sites*; ES/ER/TM-200; U.S. Department of Energy: Washington, DC, 1996.

Suter II, G.W., B.W. Cornaby, C.T. Hadden, R.N. Hull, M. Stack, and F.A. Zafran. An approach for balancing health and ecological risks at hazardous waste sites. *Risk Anal.* **1995**, *15*, 221–231.

Swanson, S.M., and R.N. Hull. Ecological Risk Assessment–Weight of Evidence Approach. Draft Report Prepared for Teck Cominco Metals Ltd. Trail, British Columbia, 2003. [

Swanson, S.M., R. Schryer, R. Shelast, P. Kloepper-Sams, and J.W. Owens. 1994. Exposure of fish to biologically treated bleached-kraft mill effluent. 3. Fish habitat and population assessment. *Environ. Toxicol. Chem.* **1994**, *13*, 1497–1507.

Swanson, S.M., C.G.J. Fraikin, D.J. Van Den Assem, J.M. Culp, R.B. Lowell, S. Eipert, G.A. Lamberti, and B.R. Taylor. *Multiple Stressors: Literature Review and Gap Analysis;* Water Environment Research Foundation: Alexandria, VA, 2004.

Swift, M.C., N.H. Troelstrup Jr., N.E. Detenbeck, and J.L Foley. Large artificial streams in toxicological and ecological research. *J. N. Am. Benthol. Soc.* **1993**, *12*, 359–366.

Touart, L. Regulatory endpoints and the experimental design of aquatic mesocosm tests. In *Aquatic Mesocosm Studies in Ecological Risk Assessment*, F.L. Graney, J.J. Kennedy, and J.H. Rogers, Eds.; SETAC Press: Pensacola, FL, 1994; pp. 25–34.

U.S. EPA (U.S. Environmental Protection Agency). *Stressor Identification Guidance Document;* EPA-822-B-00-025; Office of Water, Office of Research and Development: Washington, DC, 2000.

U.S. EPA. *Guidelines for Ecological Risk Assessment;* EPA/630/R-95/002F; Risk Assessment Forum, Washington, DC, April 1998.

U.S. EPA. *The 1996 Clean Water Needs Survey*; EPA/832/F-99/059; U.S. Environmental Protection Agency: Washington, DC, 1996.

U.S. EPA. *Framework for Ecological Risk Assessment;* EPA/630/R-92/001; Risk Assessment Forum, Washington, DC, 1992.

Vollenweider, R.A. *A Manual on Methods for Measuring Primary Production in Aquatic Environments*; International Biological Programme Handbook No. 12; Blackwell Scientific: Oxford, U.K., 1969.

Wang, W.X., and R.C.H. Dei. Effects of major nutrient additions on metal uptake in phytoplankton. *Environ. Pollut.* **2001**, *111*, 233–240.

Ward, A.K., and M.D. Johnson. Heterotrophic microorganisms. In *Methods in Stream Ecology,* Hauer, F.R., and Lamberti, G.A. Eds.; Academic Press: New York, 1996; pp. 233–268.

Washington State Department of Ecology. *Sediment Management Standards*; Washington State Department of Ecology: Olympia, WA, 1991; Chap. 173–204 WAC.

Gerritsen, J., A. Richardson, and C. Victoria. *Critical Research Needs for Understanding Ecosystem Risks from Multiple Stressors*; WERF 97-IRM-3; Water Environment Research Foundation: Alexandria, VA, 1998.

Webber, E.C., W.G. Deutsch, D.R. Bayne, and W.C. Seesock. Ecosystem level testing of a synthetic pyrethroid insecticide in aquatic mesocosms. *Environ. Toxicol. Chem.* **1992**, *11*, 87–105.

White, D.A., A. Engelmann, D. Williams, and M. Smith. *The Relationship between Watershed Stressors, Riparian Structure and In-Stream Biotic Integrity for Selected Watersheds on the Eastern Edge of the Cornbelt. 1998.* Paper presented at the SETAC Workshop on Modeling and Measuring the Vulnerability of Ecosystems at Regional Scales for Use in Ecological Risk Assessment and Risk Management; Seattle, August, 1998.

White-Hull, C.E., S.D. Dyer, C.A. Pittinger, and C.C. Kuta. *Critique of U.S. EPAís Index of Watershed Indicators. 1998.* Paper presented at the SETAC Workshop on Modeling and Measuring the Vulnerability of Ecosystems at Regional Scales for Use in Ecological Risk Assessment and Risk Management; Seattle, August, 1998

Yoder, C.O., and E.T. Rankin. Biological response signatures and the area of degradation value: new tools for interpreting multimetric data, In *Biological Assessment and Criteria: Tools for Water Resource Planning and Decision Making*, W.S. Davis and T. Somon, Eds.; Lewis Publishers: Boca Raton, FL, 1995; Chap. 17, pp. 263–286.

WERF SUBSCRIBERS

WASTEWATER UTILITY

Alabama
Montgomery Water Works & Sanitary Sewer Board

Alaska
Anchorage Water & Wastewater Utility

Arizona
Gila Resources

Glendale, City of, Utilities Department

Mesa, City of

Peoria, City of

Pima County Wastewater Management

Arkansas
Little Rock Wastewater Utility

California
Calaveras County Water District

Central Contra Costa Sanitary District

Contra Costa Water District

Crestline Sanitation District

Delta Diablo Sanitation District

Dublin San Ramon Services District

East Bay Dischargers Authority

East Bay Municipal Utility District

El Dorado Irrigation District

Fairfield-Suisun Sewer District

Fresno Department of Public Utilities

Irvine Ranch Water District

Las Virgenes Municipal Water District

Livermore, City of

Lodi, City of

Los Angeles, City of

Los Angeles County, Sanitation Districts of

Napa Sanitation District

Orange County Sanitation District

Palo Alto, City of

Riverside, City of

Sacramento Regional County Sanitation District

San Diego Metropolitan Wastewater Department, City of

San Francisco, City & County of

San Jose, City of

Santa Barbara, City of

Santa Rosa, City of

South Bayside System Authority

South Coast Water District

South Orange County Wastewater Authority

Sunnyvale, City of

Union Sanitary District

West Valley Sanitation District

Colorado
Boulder, City of

Colorado Springs Utilities

Littleton/Englewood Water Pollution Control Plant

Metro Wastewater Reclamation District, Denver

Connecticut
The Mattabasset District

New Haven, City of, WPCA

District of Columbia
District of Columbia Water & Sewer Authority

Florida
Broward, County of

Fort Lauderdale, City of

Gainesville Regional Utilities

JEA

Miami-Dade Water & Sewer Authority

Orange County Utilities Department

Orlando, City of

Reedy Creek Improvement District

Seminole County Environmental Services

St. Petersburg, City of

Stuart Public Utilities

Tallahassee, City of

Tampa, City of

Toho Water Authority

West Palm Beach, City of

Georgia
Atlanta Department of Watershed Management

Augusta, City of

Clayton County Water Authority

Cobb County Water System

Columbus Water Works

Fulton County

Gwinnett County Department of Public Utilities

Savannah, City of

Hawaii
Honolulu, City and County of

Illinois
American Bottoms Wastewater Treatment Plant

Greater Peoria Sanitary District

Kankakee River Metropolitan Agency

Metropolitan Water Reclamation District of Greater Chicago

North Shore Sanitary District

Wheaton Sanitary District

Iowa
Ames, City of

Cedar Rapids Wastewater Facility

Des Moines Metro Wastewater Reclamation Authority

Iowa City, City of

Kansas
Johnson County Unified Wastewater Districts

Unified Government of Wyandotte County/Kansas City, City of

Kentucky
Louisville & Jefferson County Metropolitan Sewer District

Louisiana
Sewerage & Water Board of New Orleans

Maine
Bangor, City of

Portland Water District

Maryland
Anne Arundel County Bureau of Utility Operations

Howard County Department of Public Works

Washington Suburban Sanitary Commission

Massachusetts
Boston Water & Sewer Commission

Upper Blackstone Water Pollution Abatement District

Michigan
Ann Arbor, City of

Detroit, City of

Holland Board of Public Works

Lansing, City of

Owosso Mid-Shiawassee County WWTP

Saginaw, City of

Wayne County Department of Environment

Wyoming, City of

Minnesota
Rochester, City of

Western Lake Superior Sanitary District

Missouri
Independence, City of

Kansas City Missouri Water Services Department

Little Blue Valley Sewer District

Nebraska
Lincoln Wastewater System

Nevada
Henderson, City of

New Jersey
Bergen County Utilities Authority

Passaic Valley Sewerage Commissioners

New York
New York City Department of Environmental Protection

Rockland County Solid Waste Management Authority/ Sewer District

North Carolina
Charlotte/Mecklenburg Utilities

Durham, City of

Metropolitan Sewerage District of Buncombe County

Orange Water & Sewer Authority

Ohio
Akron, City of

Butler County Department of Environmental Services

Columbus, City of

Metropolitan Sewer District of Greater Cincinnati

Northeast Ohio Regional Sewer District

Summit, County of

Oklahoma
Tulsa, City of

Oregon
Clean Water Services

Eugene/Springfield Water Pollution Control

Water Environment Services

Pennsylvania
Philadelphia, City of

University Area Joint Authority, State College

South Carolina
Charleston Commissioners of Public Works

Mount Pleasant Waterworks & Sewer Commission

Spartanburg Sanitary Sewer District

Tennessee
Cleveland, City of

Knoxville Utilities Board

Murfreesboro Water & Sewer Department

Nashville Metro Water Services

Texas
Austin, City of

Dallas Water Utilities

Denton, City of

El Paso Water Utilities

Fort Worth, City of

Gulf Coast Waste Disposal Authority

Houston, City of

San Antonio Water System

Trinity River Authority

Utah
Salt Lake City Corporation

Virginia
Alexandria Sanitation Authority

Arlington, County of

Fairfax County Virginia

Hampton Roads Sanitation District

Henrico, County of

Hopewell Regional Wastewater Treatment Facility

Loudoun County Sanitation Authority

Lynchburg Regional WWTP

Prince William County Service Authority

Richmond, City of

Rivanna Water & Sewer Authority

Washington
Edmonds, City of

Everett, City of

King County Department of Natural Resources

Seattle Public Utilities

Sunnyside, Port of

Yakima, City of

Wisconsin
Green Bay Metro Sewerage District

Madison Metropolitan Sewerage District

Milwaukee Metropolitan Sewerage District

Racine, City of

Sheboygan Regional Wastewater Treatment

Wausau Water Works

Australia
South Australian Water Corp.

Sydney Water Corp.

Water Corp. of Western Australia

Canada
Greater Vancouver Regional District

Toronto, City of, Ontario

Winnipeg, City of, Manitoba

Mexico
Servicios de Agua y Drenaje de Monterrey, I.P.D.

New Zealand
Watercare Services Limited

Singapore
Singapore Public Utilities Board

United Kingdom
Yorkshire Water Services Limited

STORMWATER UTILITY

California
Los Angeles, City of, Department of Public Works

Monterey, City of

San Francisco, City & County of

Santa Rosa, City of

Sunnyvale, City of

Colorado
Boulder, City of

Georgia
Griffin, City of

Iowa
Cedar Rapids Wastewater Facility

Des Moines Metro Wastewater Reclamation Authority

Kansas
Overland Park, City of

Kentucky
Louisville & Jefferson County Metropolitan Sewer District

Maine
Portland Water District

Minnesota
Western Lake Superior Sanitary District

North Carolina
Charlotte, City of, Stormwater Services

Pennsylvania
Philadelphia, City of

Tennessee
Chattanooga Storm Water Management

Washington
Seattle Public Utilities

STATE

Arkansas Department of Environmental Quality

Fresno Metropolitan Flood Control District, Calif.

Urban Drainage & Flood Control District, Colo.

CORPORATE

ADS Environmental Services

The ADVENT Group Inc.

Alan Plummer & Associates

Alden Research Laboratory

Aqua-Aerobic Systems, Inc.

Aquateam–Norwegian Water Technology Centre A/S

BaySaver Inc.

BioVir Laboratories, Inc.

Black & Veatch

Boyle Engineering Corporation

Brown & Caldwell

Burns & McDonnell

CABE Associates Inc.

The Cadmus Group

Camp Dresser & McKee Inc.

Carollo Engineers Inc.

Carpenter Environmental Associates Inc.

CDS Technologies Inc.

Chemtrac Systems Inc.

CH2M HILL

Damon S. Williams Associates, LLC

David L. Sheridan, P.C.

Dewling Associates, Inc.

Earth Tech Inc.

Eco-Matrix

Ecovation

Environmental Engineers International

EMA Inc.

The Eshelman Company Inc.

Finkbeiner, Pettis, & Strout Inc.

Freese & Nichols, Inc.

ftn Associates Inc.

Fuss & O'Neill, Inc.

Gannett Fleming Inc.

Golder Associates Inc.

Greeley and Hansen LLC

Hazen & Sawyer, P.C.

HDR Engineering Inc.

HNTB Corporation

HydroQual Inc.

Infilco Degremont Inc.

Ingersoll-Rand Energy Systems

Insituform Technologies Inc.

Institute for Environmental Technology & Industry, Korea

Jacobson Helgoth Consultants Inc.

Jason Consultants Inc.

Jordan, Jones, & Goulding Inc.

KCI Technologies Inc.

Kelly & Weaver, P.C.

Kennedy/Jenks Consultants

Komline Sanderson Engineering Corporation

Lawler, Matusky & Skelly Engineers, LLP

Limno-Tech Inc.

Lombardo Associates Inc.

Malcolm Pirnie Inc.

McKim & Creed

MEC Analytical Systems Inc.

Metcalf & Eddy Inc.

MWH

New England Organics

Odor & Corrosion Technology Consultants Inc. (OCTC)

Oswald Green, LLC

PA Government Services Inc.

Parametrix Inc.

Parsons

Post, Buckley, Schuh & Jernigan

R&D Engineering/Conestoga Rover & Associates

The RETEC Group

RMC, Inc. (Raines, Melton & Carella)

R.M. Towill Corporation

Ross & Associates Ltd.

Royce Technologies

SAIC Maritime Technical Group

Short Elliott Hendrickson, Inc.

Stantec Consulting Group, Inc.

Stormwater Management, Inc.

Synagro Technologies, Inc.

Tetra Tech Inc.

Trojan Technologies Inc.

URS Corporation

Veolia Water NATC

Wade-Trim Inc.

Weston Solutions Inc.

Woodard & Curran

WRc/D&B, LLC

WWETCO, LLC

Zoeller Pump Company

INDUSTRY

American Electric Power

ChevronTexaco Energy Research & Technology Company

The Coca-Cola Company

Dow Chemical Company

DuPont Company

Eastman Kodak Company

Eli Lilly & Company

Merck & Company Inc.

ONDEO Services

Procter & Gamble Company

PSEG Services Corp.

RWE Thames Water Plc

Severn Trent Services Inc.

United Water Services LLC

Note: List as of 7/1/04

WERF Product Order Form

As a benefit of joining the Water Environment Research Foundation, subscribers are entitled to receive one complimentary copy of all final reports and other products.Additional copies are available at cost (usually $10).To order your complimentary copy of a report, please write "free" in the unit price column.WERF keeps track of all orders.If the charge differs from what is shown here, we will call to confirm the total before processing.

Name _______________ Title _______________

Organization _______________

Address _______________

City _______________ State _______ Zip Code _______ Country _______

Phone _______________ Fax _______________ Email _______________

Stock #	Product	Quantity	Unit Price	Total
			Postage & Handling	
			VA Residents Add 4.5% Sales Tax	
			Canadian Residents Add 7% GST	
			TOTAL	

Method of Payment: (All orders must be prepaid.)

❑ Check or Money Order Enclosed

❑ Visa ❑ Mastercard ❑ American Express

Account No. _______________ Exp. Date _______

Signature _______________

Shipping & Handling:

Amount of Order	United States	Canada & Mexico	All Others
Up to but not more than:	Add:	Add:	Add:
$20.00	$5.00*	$8.00	50% of amount
30.00	5.50	8.00	40% of amount
40.00	6.00	8.00	
50.00	6.50	14.00	
60.00	7.00	14.00	
80.00	8.00	14.00	
100.00	10.00	21.00	
150.00	12.50	28.00	
200.00	15.00	35.00	
More than $200.00	Add 20% of order	Add 20% of order	

*minimum amount for all orders

Note: Please make checks payable to the Water Environment Research Foundation.

To Order (Subscribers Only):

Log on to www.werf.org and click on the "Product Catalog."

Phone: (703) 684-2470
Fax: (703) 299-0742.

WERF
Attn: Subscriber Services
635 Slaters Lane
Alexandria, VA 22314-1177

To Order (Non-Subscribers):

Non-subscribers may be able to order WERF publications either through WEF (www.wef.org) or IWAP (www.iwapublishing.com).Visit WERF's website at www.werf.org for details.

IWA Publishing's authorised EU representative for General Product Safety Regulations is Diane D'Arras, 15 rue Duret, 75116 Paris, France, e-mail: safety@iwap.co.uk.

Printed and bound by CPI Group (UK) Ltd, Croydon, CR0 4YY
16/06/2026
02140375-0001